Encounters with Einstein

ENCOUNTERS WITH EINSTEIN

And Other Essays on People, Places, and Particles

Werner Heisenberg

Princeton University Press
Princeton, New Jersey

Published by Princeton University Press, 41 William Street, Princeton, New Jersey 08540

Copyright © 1983 by Werner Heisenberg

All Rights Reserved

This book was originally published by Seabury Press in 1983 as *Tradition in Science*, and is reprinted in the Princeton Science Library edition by arrangement with Harper & Row

Heisenberg, Werner, 1901-1976.
 [Tradition in der Weissenschaft, English]
 Encounters with Einstein : and other essays on people, places, and particles/Werner Heisenberg.
   p. cm.
 Translation of: Tradition in der Wissenschaft.
 Reprint. Originally published: Tradition in science. San Francisco, Calif. : Seabury Press, 1983.
 ISBN 0-691-02433-2 (alk. paper)
 1. Science—History. 2. Physics—History. 3. Particles (Nuclear physics) 4. Quantum theory—History. I. Title.
Q175.H393113   1989
509—dc20
               89-33075

First Princeton Paperback printing, 1989

9 8 7 6 5 4

Princeton University Press books are printed on acid-free paper and meet the guidelines for permanence and durability of the Committee on Production Guidelines for Book Longevity of the Council on Library Resources

Printed in the United States of America

http://pup.princeton.edu

Contents

Tradition in Science	1
Development of Concepts in the History of Quantum Mechanics	19
The Beginnings of Quantum Mechanics in Göttingen	37
Cosmic Radiation and Fundamental Problems in Physics	56
What is an Elementary Particle?	71
The Role of Elementary Particle Physics in the Present Development of Science	89
Encounters and Conversations with Albert Einstein	107
The Correctness-Criteria for Closed Theories in Physics	123
Thoughts on The Artist's Journey into the Interior	130
Epilogue, by Hans-Peter Dürr	136

Tradition in Science *

As we celebrate the five hundredth birthday of Copernicus, we believe that our present science is connected with his work; that the direction which he had chosen for his research in astronomy still determines to some extent the scientific work of our time. We are convinced that our present problems, our methods, our scientific concepts are, at least partly, the results of a scientific tradition which accompanies or leads the way of science through the centuries. It is therefore natural to ask to what extent our present work is determined or influenced by tradition. Are the *problems* in which we are engaged freely chosen according to our interest or inclination, or are they given to us by an historical process? To what extent can we select our scientific *methods* according to the purpose, to what extent do

---

*From *Science and Public Affairs—Bulletin of the Atomic Scientists*, 29, No. 10, pp. 4–10 (1973); and in *The Nature of Scientific Discovery*, ed. O. Gingerich, Smithsonian Institution Press, 1975. (Lecture given on April 24, 1973 in Washington at the Symposium of the Smithsonian Institution and the National Academy of Sciences; written in English.)

we again follow a given tradition? And finally how free are we in choosing the *concepts* for formulating our questions? Any scientific work can only be defined by formulating the questions which we want to answer. But in order to formulate the questions we need concepts by which we hope to get hold of the phenomena. These concepts usually are taken from the past history of science; they suggest already a possible picture of the phenomena. But if we are going to enter into a new realm of phenomena, these concepts may act as a collection of prejudices, which hamper progress rather than foster it. Even then we have to use concepts, and we can't help falling back on those given to us by tradition. Therefore I will try to discuss the influence of tradition first in the selection of problems, then in the scientific methods and finally in the use of concepts as tools for our work.

To what extent are we bound by tradition in the selection of our problems? When we look back into the history of science, we see that periods of intense activity alternate with long periods of inactivity. In ancient Greece the philosophers started asking questions of principle with respect to the phenomena in nature. There had been a considerable practical knowledge long before; great skill had been developed in building houses, cutting and moving big stones, constructing ships and so on; but it was first in the period after Pythagoras, that this skill was supplemented by scientific inquiry. The relevance of mathematical relations in natural phenomena was discovered by Pythagoras and his pupils, and a great development both in mathematics, in astronomy and in natural philosophy followed. The decline of Greek science after the Hellenistic period, after Ptolemy the last great astronomer, marked the beginning of a long period of inactivity which lasted until the Renaissance in Italy. During this period of stagnation again an admirable development of practical knowledge led to a high civilization in the Arab countries; but it was not accompanied by a corresponding development in science, by a deeper understanding of nature. More than a thousand years later, when humanism and renaissance had shown the way to a more liberal trend of thought, when the explorers had demonstrated the possibility of expansion on our earth, then a new activity in science was inaugurated by the discoveries of Copernicus, Galileo and Kepler. This activity has lasted until our present time, and we do not know whether it will still continue for long, or will give way to a new period in which the interest goes off into very different directions.

Looking back upon history in this way, we see that we apparently have little freedom in the selection of our problems. We are bound up with the historical process, our lives are parts of this process, and our choice seems to be restricted to the decision whether or not we want to participate in a development that takes place in our time, with or without our contribution. Without such a favorable development, our activity would probably be lost. If Einstein had lived in the twelfth century, he would have had very little chance to become a good scientist. And even within such a fruitful period the scientist has not much choice in selecting his problems. On the contrary, one may say that a fruitful period is characterized by the fact that the problems are given, that we need not invent them. This seems to be true in science as well as in art. When, in the fifteenth century, the painters in the Netherlands had discovered the possibility of portraying men as active members of their society, many gifted people were attracted by this possibility and competed there in solving this given problem. In the eighteenth century Haydn had tried, in his string quartets, to express those emotions that had become visible in the literature of his time, in the work of Rousseau and in Goethe's *Werther;* and then the musicians of the younger generation, Mozart, Beethoven, Schubert, gathered in Vienna competing in the solution of this problem. In our century the development of physics had led Niels Bohr to the idea that

Rutherford's experiments on α-rays, Planck's theory of radiation, and the facts of chemistry, could be combined into a theory of the atom; and in the following years many young physicists went to Copenhagen in order to participate in the solution of this given problem. One cannot doubt that in the selection of problems the tradition, the historical development, plays an essential role.

This may also sometimes be true in a negative sense. It can happen that traditional themes have been exhausted and that the gifted people turn away from a field in which they see no more objects for their activity. After Thomas Aquinas, the philosophers got tired of the theological and philosophical problems of the scholastics, and turned to humanism. In our time the traditional themes of art seem to be exhausted. Last year one of the most popular yearly exhibitions of modern art in Germany, called "Documenta," in Kassel, was a center rather of political propaganda than of art, and on the outside of the exhibition building the young artists had fixed a huge poster with the text: "art is superfluous." In a similar way we cannot exclude the possibility that, after some time, the themes of science and technology will be exhausted, that a younger generation will be tired of our rationalistic and pragmatic attitude and will turn their interest to an entirely different activity. In the present situation, however, many problems are still given in pure and in applied science, no effort is needed to invent them, and they will be passed on from the teachers to their pupils.

In this connection it is important to emphasize the very great role of personal relations in the development of science or art. It need not only be the relation between teacher and pupil, it may simply be personal friendship or respect between people working for the same goal. This is probably the most efficient instrument of tradition. Among the many examples which could be mentioned for this kind of tradition I will only recall some of the personal relations which have shaped the history of physics in the first half of our present century. Einstein was well acquainted with Planck, he corresponded with Sommerfeld about the theory of relativity and about quantum theory, he was a close friend of Max Born, though he could never agree with him on the statistical interpretation of quantum theory, and he discussed with Niels Bohr the philosophical implications of the relations of uncertainty. A large part of the scientific analysis of those extremely difficult problems arising out of relativity and quantum theory was actually carried out in conversations between those who took an active part in the research.

Sommerfeld's school in Munich was a center of research in the early twenties; Pauli, Wentzel, Laporte, Lenz and many others belonged to this group, and we discussed almost daily the difficulties and paradoxes in the interpretation of recent experiments. When Sommerfeld had received a letter from Einstein, or Bohr, he read the important parts of the letter in the seminar and started at once a discussion on the critical problems. Niels Bohr held close connection with Lord Rutherford, Otto Hahn and Lise Meitner, and he considered the continuous exchange of information between experiment and theory as a central task in the progress of physics. The enormous influence of Niels Bohr on the development of physics in his time was not primarily due to his papers, but to his way of discussing again and again with his partners the fundamental difficulties of quantum theory—which, as he knew, did not allow of any cheap solution. When wave mechanics was introduced by Schrödinger, Bohr saw at once that this was a very important new aspect of quantum theory, but that a simple replacement of the electronic orbits in the atom by three-dimensional matter waves could not solve the real difficulties. Again the only way of analyzing the problem seemed to be a personal discussion with the author. Schrödinger was invited to Copenhagen, and in two weeks of most intense discussions the way was prepared for the later development in the interpretation of quantum theory, for the concept of complementarity and the relations of uncertainty. I need not enlarge upon these examples. It is obvious that personal relations play a decisive role in the progress of science and in the selection of problems.

There are of course also other motives for the selection of problems, which have played their role in the history of science. The best known of these motives is the practical applicability of science. Already in ancient times the interest in astronomy and mathematics was stimulated by the fact that knowledge in these fields was helpful for navigation and for the surveying of land. Navigation played a very important role in the fifteenth century, when the explorers left Europe and the Mediterranean and went westward. It is certainly no mere accident that the discovery of Copernicus happened shortly after the beginning of this period. When Galileo defended the ideas of Copernicus, he made use of a newly invented instrument, the telescope, thereby demonstrating, that the practical tool may be helpful in the progress of science, and that science may be helpful in leading to the invention of practical tools. Galileo and his followers were strongly interested in the practical side of science. They studied mechanical devices, e.g., the mechanical clock; they invented optical instruments; Newton constructed a bridge across the river Cam in Cambridge, and so on. It has always belonged to the tradition in science, guiding the activity of many generations, that science should be applied to practical purposes and that practical application should be a check on the validity of the results and a justification for the efforts in science. Hence the atomic physicists of the first half of our present century were merely following this old tradition of science when they looked for practical applications of atomic physics. It was of course extremely disappointing for them that the first practical application was for warfare. Still, the fact that one could now transmute chemical elements into others in large quantities was justly considered as a real triumph of science.

This interest in the practical application of science is frequently misunderstood as the trivial attempt of the scientist to acquire economic wealth, to earn money. It is true that this trivial motive does play a role sometimes, depending of course on the individual people. But it should not be overestimated. There is another much stronger motive which fascinates the good scientist in connection with practical application, namely: to see that it works; to see that one has correctly understood nature. I remember a conversation with Enrico Fermi after the war, a short time before the first hydrogen bomb was to be tested in the Pacific. We discussed this plan and I suggested that one should perhaps abstain from such a test, considering the biological and political consequences. Fermi replied, "But it is such a beautiful experiment." This is probably the strongest motive behind the applications of science; the scientist needs the confirmation from an impartial judge, from nature herself, that he has understood her structure. And he wants to see the effect of his effort.

From this attitude one can also easily understand the motives which determine the line of research for the individual scientist. Such a line of research is usually based on some theoretical ideas, on conjectures concerning the interpretation of the known phenomena. But which theory is accepted? History teaches that it is usually not the consistency, the clarity of a theory which makes it acceptable, but the hope that one can participate in its elaboration and its verification. It is the wish for our own activity, the hope for results of our own efforts, which leads us on our way through science. This wish is stronger than our rational judgment about the merits of various theoretical ideas. In the early twenties we knew that Bohr's theory of the atom could not possibly be quite correct. But we guessed that it pointed in the right direction, and we hoped that we would be able some day to avoid the inconsistencies and to replace Bohr's theory by a more satisfactory picture.

But the role of tradition in science is not restricted to the selection of problems, and thereby I am coming to the second part of my talk. Tradition exerts its full influence in deeper layers of the scientific process, where it is not so easily visible; and there we should first of all mention the scientific method. In the scientific work of our present century, we still follow essentially the method discovered and developed by Copernicus, Galileo and their successors in the sixteenth and seventeenth centuries. This method is sometimes misunderstood by terming it "empirical science," as contrasted to the speculative science of former centuries. Actually Galileo turned away from the traditional science of his time, which was based on Aristotle, and took up the philosophical ideas of Plato. He replaced the descriptive science of Aristotle by the structural science of Plato. When he argued for experience he meant experience illuminated by mathematical constructs. Galileo, as well as Copernicus, had understood that by going away from immediate experience, by idealizing experience, we may discover mathematical structures in the phenomena, and thereby gain a new simplicity as a basis for a new understanding. Aristotle, for example, had correctly stated that light bodies fall more slowly than heavy bodies. Galileo claimed that all bodies fall with the same speed in empty space, and that their fall can be described by simple mathematical laws. Fall in empty space could not be observed accurately in his time; but Galileo's claim suggested new experiments. The new method did not aim at the description of what is visible, but rather at the design of experiments and the production of phenomena that one does not normally see, and at their calculation on the basis of mathematical theory.

Therefore two features are essential for the new method: the attempt to design new and very accurate experiments which idealize and isolate experience, and thereby actually create new phenomena, and the comparison of these phenomena with mathematical constructs, called natural laws. Before we discuss the validity of this method even in our present science, we should perhaps briefly ask for the basis of confidence, which led Copernicus, Galileo and Kepler on this new way. Following a paper of von Weizsäcker, I think we have to state that this basis was mainly theological. Galileo argued that nature, God's second book (the first one being the Bible) is written in mathematical letters, and that we have to learn this alphabet if we want to read it. Kepler is even more explicit in his work on world harmony; he says: God created the world in accordance with his ideas of creation. These ideas are the pure archetypal forms which Plato termed Ideas, and they can be understood by Man as mathematical constructs. They can be understood by Man, because Man was created as the spiritual image of God. Physics is reflection on the divine Ideas of Creation, therefore physics is divine service.

We are in our time very far from this theological foundation or justification of physics; but we still follow this method, because it has been so successful. The essential basis for the success was the possibility of repeating the experiments. We can finally agree about their results, because we have learned that experiments carried out under precisely the same conditions do actually lead to the same results. This is not at all obvious. It can only be true if the events follow exactly a causal chain, a sequence of cause and effect. But on account of its success this kind of causality has—in the course of years—been accepted as one of the fundamental principles of science. The philosopher Kant has stressed the point, that causality in this sense is not an empirical law, but belongs to our method of science; it is the condition for that kind of science which was inaugurated in the sixteenth century and has been elaborated ever since.

A consequence of this attitude in science is the assumption that we study nature as it "really is." We imagine a world which exists in space and time and follows its natural laws, independent of any observing subject. Therefore in observing the phenomena, we take great care to eliminate any influence from the observer. When we produce new phenomena by means of our experimental equipment we are convinced that we do not really produce new phenomena; that actually these phenomena occur frequently in nature without our interference, and that our equipment is just made to isolate and to study them. In all these points we still follow confidently the tradition from the time of Copernicus and Galileo.

But are we really entitled to do so, considering the well-known epistemological difficulties of quantum theory? In the big accelerators, for example, we study the collisions between elementary particles, and we imagine that, even if we had not built the accelerators, such phenomena would occur in our atmosphere on account of the cosmic radiation. But would there be waves or particles coming from the outside, would they produce interference patterns or tracks: What does actually happen when we do not observe, and do we know what the word "actually" means in this context? These are hard questions, and we see that tradition can lead us into difficulties.

It is generally believed that our science is empirical, and that we draw our concepts and our mathematical constructs from the empirical data. If this were the whole truth, we should, when entering into a new field, introduce only such quantities as can directly be observed, and formulate natural laws only by means of these quantities. When I was a young man, I believed that this was just the philosophy which Einstein had followed in his theory of relativity. Therefore I tried to take a corresponding step in quantum theory by introducing the matrices. But when I later asked Einstein about it, he answered, "This may have been my philosophy, but it is nonsense all the same. It is never possible to introduce only observable quantities in a theory. It is the theory which decides what can be observed." What he meant by this remark was that, when we go from the immediate observation—a black line on a photographic plate, a discharge in a counter or suchlike—to the phenomena we are interested in, we must make use of theory and of theoretical concepts. We cannot separate the empirical process of observation from the mathematical construct and its concepts. The most conspicuous demonstration of this thesis of Einstein's was the later discovery of the relations of uncertainty.

But this new situation in quantum theory does not necessarily question the traditional method in science; it only questions the assumption that concepts and mathematical constructs can simply be taken from experience. It is true that in quantum theory we cannot rely on strict causality. But by repeating the experiments many times we can finally derive from the observations statistical distributions, and by repeating such series of experiments we can arrive at objective statements concerning these distributions. This is a standard method in particle physics which may be considered as a natural extension of the traditional method.

So finally it seems that with regard to the scientific method we follow strictly the tradition inaugurated in the time of Galileo. In spite of the many different fields which have been developed—physics, chemistry, biology, atomic and nuclear science etc.—the fundamental method has always been the same. One has the impression that in this period most scientists believed that this was the only acceptable method, which could lead to objective, and that meant to correct statements, concerning the behavior of nature.

There has been one attempt to work on an entirely different line which I should mention. The German poet Goethe tried to return to a descriptive science; a science which is interested only in the visible natural phenomena, not in experiments which produce artificial new effects. He objected to the separation of the phenomena into their objective and their subjective sides, and he was filled with fear of the destruction of nature by an overflowing of technical science. In our time, when we know of the contamination of air and water, the poisoning of the soil by chemical fertilizers, the atomic weapons, we understand Goethe's fear better than his contemporaries could. But Goethe's attempt did not really influence the course of science. The success of the traditional method was too overwhelming.

In addition to the effect of tradition in the selection of problems, and in the scientific method, the influence of tradition is perhaps strongest in shaping or passing on the concepts by which we try to get hold of the phenomena. The history of science is not only a history of discoveries and observations, it is also a history of concepts, and therefore, in this third part of my talk, I will try to discuss briefly the history of concepts during the period following Copernicus and Galileo, and the role of tradition in this history.

The new science started with astronomy, and therefore the positions and the velocities of bodies were natural first concepts for describing the phenomena. Newton, in his "mathematical principles of natural philosophy" added the concepts of mass and of force; he introduced the "quantity of motion," which is essentially what we call momentum, and later such concepts as kinetic and potential energy completed the conceptual basis of mechanics. It remained for more than a century the basis of exact science as a whole, and its success was so convincing that, whenever the phenomena suggested new concepts, the scientists tried to follow the tradition and to reduce them to the old ones. The motion of fluids was pictured as motion of the infinitely many smallest parts of the fluid, their dynamical behavior was successfully treated according to Newton's laws. When, in the second half of the eighteenth century, the interest was concentrated on electricity and magnetism, the concept of force was used for describing the phenomena; and force was meant in the sense of mechanics, a force acting instantaneously and depending only on the positions and the velocities of the bodies concerned. For understanding the different states and the chemical behavior of matter, Gassendi had revived the idea of its atomic constitution, and his followers used Newtonian mechanics for describing the motion of the atoms and the resulting properties of matter. A beam of light could be considered as consisting either of small quickly moving particles, or of waves. But even the waves would be the waves in some kind of material, and one could hope that finally the smallest parts of this material could be treated according to Newton's laws.

As in the case of the scientific methods, nobody doubted that this reduction to the mechanical concepts could finally be effected. But here history decided otherwise. In the nineteenth century it became clear, gradually, that the electromagnetic phenomena are of a different nature. Faraday introduced the concept of the electromagnetic field, and after the completion of the theory by Maxwell this concept gained more and more reality; the physicists began to understand that a field of force in space and time could be just as real as a position or velocity of a mass, and that there was no point in considering it as a property of some unseen material called "ether." Here tradition was more a hindrance than a help. Actually, it was not until the discovery of relativity that the idea of the ether was really given up, and thereby the hope of reducing electromagnetism to mechanics.

A similar development can be recognized in the theory of heat, but here the alienation from mechanical concepts could be seen only in rather subtle points. To begin with, everything looked very simple. A piece of matter consists of many atoms or molecules; statistical considerations about the mechanical motions of these many particles should be sufficient to describe the behavior of matter under the influence of heat or chemical changes. The concepts of temperature and of entropy seemed just adequate to get hold of this statistical behavior. I think it was Willard Gibbs who first understood what an abyss had been opened up in physics by these concepts. His idea of the canonical ensemble demonstrates that the word *temperature* characterizes our degree of knowledge of the mechanical behavior of the atoms, but not the objective mechanical behavior. The word refers to a certain kind of observation, in that it requires an exchange of heat between the system and the measuring equipment, the thermometer; it requires a thermodynamical equilibrium. Therefore, if we know the temperature of a system, we cannot know its energy accurately, the inaccuracy depending on the number of degrees of freedom in the system. Of course, tradition worked very strongly against this kind of interpretation, and I believe that the majority of physicists did not accept it until finally, in our century, quantum theory was completed. But I would like to mention that, when I entered Niels Bohr's institute in Copenhagen in 1924, the first thing Bohr demanded was that I should read Gibbs' book on thermodynamics. And he added that Gibbs had been the only physicist who had really understood statistical thermodynamics.

But worse things happened in other fields. In the theory of relativity and in quantum theory we had to learn that some of the oldest traditional concepts did not work satisfactorily, and had to be replaced by better ones. Space and time are not so independent of each other as Newton had believed; they are related by the Lorentz transformation. The state of a system in quantum mechanics can be characterized mathematically by a vector in a space of many dimensions, and this vector implies statements concerning the statistical behavior of the system under given conditions of observation. An objective description of the system in the traditional sense is impossible. I need not go into the details. You know how difficult it has been for the physicists to accept these changes in the fundamental concepts.

Since it is my task to speak about the role of tradition in science, I have to ask whether tradition has really been only a hindrance in these developments, whether it has merely filled the minds of scientists with prejudices, the removal of which was the most important condition for progress. At this point the problem arises from the word *prejudice*. When we speak about our investigations, about the phenomena we are going to study, we need a language, we need words, and the words are the verbal expression of concepts. In the beginning of our investigations, there can be no avoiding the fact, that the words are connected with the old concepts, since the new ones don't yet exist. Therefore these so-called prejudices are a necessary part of our language, and cannot simply be eliminated. We learn language by tradition, the traditional concepts form our way of thinking about the problems and determine our questions. When the experiments of Lord Rutherford suggested that the atom consists of a nucleus surrounded by electrons, one could not help asking: what is the location or the motion of the electrons in these outer parts of the atom? What are the electronic orbits? Or when one observed events on very distant stars, it was only sensible to ask: are these two events simultaneous or not? To realize that such questions have no meaning is a very difficult and painful process. It should not be belittled by the word *prejudice*. Therefore one may say that in a state of science where fundamental concepts have to be changed, tradition is both the condition for progress and a hindrance. Hence it usually takes a long time before the new concepts are generally accepted.

Finally, let me apply these ideas to the present state of physics. In our time the fundamental structure of matter is one of the central problems, and the concept of the elementary particle has dominated this problem since the time of Democritus. This can be clearly recognized in our pictures and our questions. A lump of matter consists of molecules; a molecule consists of atoms; an atom consists of nucleus and electrons; a nucleus consists of protons and neutrons. A proton—well, that could be an elementary particle. But we would term it "elementary" only if it could not be divided again; we would then wish it to be a point of mass and of charge. But a proton has a finite size and can be divided. From a collision between two energetic protons many pieces may emerge. But these pieces are not smaller than the proton, they are just particles like the protons; particular objects out of a whole spectrum of particles, whose charge—if it is not zero—is not smaller than that of the proton. So, what we see in such a collision should perhaps not be called a division of the proton; it is the creation of new particles out of the kinetic energy of the colliding protons. If the proton is not elementary, what does it consist of? Of matter; but matter consists of particles. Hence the proton consists of any number and any kind of particles, and so on. We see that we do not get a sensible answer to those questions, which we have asked and do ask according to the tradition, and this tradition goes back 2500 years, to the time of Democritus. But we cannot help asking these questions, since our language is bound up with this tradition. We must use words like *divide* or *consist of* or *number of particles*, and at the same time we learn from the observations, that these words have only a very limited applicability. Again it is extremely difficult to get away from the tradition. In one of the most recent papers on elementary particles I saw the statement, "From the results of Bjorkén we can conclude, that the proton in its electric properties has a granular structure." It did not occur to the author that such words as *granular structure* have perhaps no other meaning here than just the scaling law of Bjorkén; that they do not carry any further information. Or another example: many experimental physicists nowadays look for "quark" particles, particles with a charge $1/3$ or $2/3$ of the charge of the proton. I am convinced that this intense search for quarks is caused by the conscious or unconscious hope of finding the really elementary particles, the ultimate units of matter. But even if quarks could be found, from all what we know they could again be divided into two quarks and one antiquark etc.; they would not be more elementary than a proton. You see how extremely difficult it is to get away from an old tradition.

What is really needed is a change in the fundamental concepts. We will have to abandon the philosophy of Democritus and the concept of fundamental elementary particles. And we should accept instead the concept of fundamental symmetries, which is a concept derived from the philosophy of Plato. Just as Copernicus and Galileo, in their method, abandoned the descriptive science of Aristotle and turned to the structural science of Plato, so we are probably forced, in our concepts, to abandon the atomic materialism of Democritus and to turn to the ideas of symmetry in the philosophy of Plato. Again we would return to a very old tradition. But, as I said before, such changes are extremely difficult. Even with the change, many complicated details will have to be worked out, both experimentally and theoretically, in elementary particle physics; but I do not believe that there will be any spectacular break-through, apart from this change in the concepts.

After going through the three most important influences of tradition in science, those in the selection of problems, in the method, and in the concepts, I should perhaps say a few words in conclusion about the future development of science. Of course, I am not interested in futurology. But since we can scarcely work on other problems save those that are given to us by the historical process, we may ask where this process has led to new and interesting questions. In physics I would like to mention astrophysics; in this field the strange properties of the pulsars and quasars, and perhaps also the gravitational waves, can be considered as a challenge. Then there is the new and wide field of molecular biology, where concepts of very different origin, namely physical, chemical and biological concepts, meet and produce a great wealth of interesting new problems. Finally, on the practical side, we have to solve the very urgent problems posed by the deterioration of our environment. I have mentioned these points, not in order to make predictions about the future, but in order to emphasize that we need not invent our problems. The scientific tradition, i.e. the historical process, does provide us with many problems and encourages our efforts. And that is a sign of a very healthy state of affairs in science.

*Development of Concepts in the History of Quantum Mechanics**

The history of physics is not only a sequence of experimental discoveries and observations, followed by their mathematical description; it is also a history of concepts. For an understanding of the phenomena, the first condition is the introduction of adequate concepts; only with the help of the correct concepts can we really know what has been observed. When we enter a new field, very often new concepts are needed, and these new concepts usually come up in a rather unclear and undeveloped form. Later they are modified, sometimes they are almost completely abandoned and are replaced by better concepts which then, finally, are clear and well-defined. I would like to describe this development in three cases which have been important for my

---

*From *The Physicist's Conception of Nature*, ed. J. Mehra, D. Reidel Publishing Co., Dordrecht, Boston, 1973, pp. 264–75. (Lecture at the Symposium on the Development of the Physicist's Conception of Nature in the Twentieth Century, held at the International Centre for Theoretical Physics, Miramare, Trieste, Italy, September 18–25, 1972; written in English.)

own work. First, the concept of the discrete stationary state, which obviously is a fundamental concept in quantum theory. Then, the concept of state, not necessarily stationary or discrete, which could only be understood after quantum mechanics and wave mechanics had been developed. And finally, closely connected with the first two, the concept of the elementary particle, which is under discussion to this day. So the first two parts of my talk will be historical, though I do not intend to recount all the mistakes and errors which we made fifty years ago—only some of them—and the last part will have to do with the problems of our present time and again with possible mistakes.

As you know, the concept of the discrete stationary state was introduced by Niels Bohr in 1913. It was the central concept in his theory of the atom, the intention of which was described thusly by Bohr: "It should be made clear that this theory is not intended to explain phenomena in the sense in which the word 'explain' has been used in earlier physics. It is intended to combine various phenomena, which seem not to be connected, and to show that they are connected." Bohr stated that only after this connection had been established could one hope to give an explanation, in the sense in which explanations were used in earlier physics. There were mainly three phenomena which had to be connected. The first was the strange fact of the stability of the atom. An atom can be perturbed by chemical processes, by collisions, by radiation or anything else, and still it always returns to its original state—its normal state. This was one fact which could not be explained satisfactorily in earlier physics. Then there were the spectral laws, especially the famous law of Ritz, that the frequency of the lines in a spectrum could be written as a difference between terms, and that these terms had to be considered as characteristic properties of the atoms. And finally there were the experiments of Rutherford, which had led him to his model of the atom.

So these three groups of facts had to be combined, and as you know, the idea of the discrete stationary state was the starting point for their combination. First of all, one had to believe that the behavior of the atom in the discrete stationary state could be explained by mechanics. This was necessary, because otherwise there would be no connection with Rutherford's model, since Rutherford's experiments were based on classical mechanics. Then also one had to combine the discrete stationary states with the frequencies of the spectrum. There one had to apply the law found by Ritz, which now was written in the form that h times the frequency of the line was equal to the difference between the energies of the initial and the final state. This law, however, could best be explained by an assumption which Bohr did not accept, namely by Einstein's idea of the light quantum. Bohr was for a long time not inclined to believe in light quanta and he therefore considered his stationary states as stations during the motion of the electron, which on its orbit around the nucleus loses energy by radiation. The assumption was that, during this process of radiation, the electron stops radiating at certain stations, called discrete stationary states. For some unknown reason it does not radiate in these stations, and the final station is the normal state of the atom. When the radiation takes place, the electron goes from one of the stationary states to the next one.

According to this picture, the time in the stationary state seemed to be longer than the time required for going from one state to another. But of course this ratio of time was never well-defined.

What could be said about radiation itself? One could use the general ideas of Maxwell's theory. From this point of view, the interaction between atom and radiation seemed to be the source of all the trouble. In the stationary state there was no such interaction, and therefore one could, so it seemed, apply classical mechanics. But could one use Maxwell's theory for radiation? I might mention that it was perhaps not necessary to take this point of view. One could have taken the light quanta more seriously. One could have said that the interference patterns which we see in light come about by some extra conditions on the motions of the light quanta. I remember vaguely a discussion with Wentzel in the old days, where he explained to me the possibility that the motion of light quanta could be quantized and thereby possibly the interference patterns could be explained. But anyway this was not the point of view which Bohr took. Wherever one started one ran into a lot of difficulties, and I would like to describe these problems in some detail.

First of all, there are strong arguments in favor of the mechanical model of the stationary state. I have mentioned the experiments of Rutherford. Then the periodic orbits of the electrons in the atom could easily be connected with quantum conditions. So the idea of the stationary state could be combined with the idea of a specified elliptical orbit of the electron. In his earlier lectures, Bohr very frequently showed pictures of electrons moving in their orbits around the nucleus.

This model worked perfectly well in a number of interesting cases. First of all, in the hydrogen spectrum. Then, in Sommerfeld's theory of the relativistic fine structure of the hydrogen lines, and in the so-called Stark effect, the splitting of lines in an electric field. So there was an enormous body of data which seemed to show that this connection of quantized electronic orbits with the discrete stationary states was correct.

On the other hand, there were other reasons for arguing that such a picture cannot be correct. I remember a conversation with Stern, who told me that in 1913, when Bohr's first paper had appeared, he had said to a friend, "If that nonsense is correct which Bohr has just published, then I will give up being a physicist."

So I now have to point out the difficulties and the errors of this model. The worst difficulty was perhaps the following. The electron described a periodic motion in the model, defined by quantum conditions, and therefore it moved around the nucleus with a certain frequency. However this frequency never turned up in the observations. You could never see it. What you saw were different frequencies, which were determined by the energy differences in the transitions from one stationary state to another. Then there was a difficulty about degeneracy. Sommerfeld had introduced the magnetic quantum number. When we have a magnetic field in some direction, then the angular momentum of the atom around this field should be one, or zero, or minus one, according to this quantum condition. But then if you take a different field, with a different direction, quantization has to be carried out with respect to this different direction. But one may have an extremely weak field, first in one direction and then, after a short time, in another direction. The field is too weak to turn the atom around. Hence the contradiction with the quantum conditions seems unavoidable.

My first discussion with Niels Bohr—which was just fifty years ago—turned upon one of these difficult points. Bohr had given a lecture in Göttingen and had stated that in a constant electric field one can calculate the energy of the stationary states according to the quantum conditions; and that a recent calculation of Kramers on the quadratic Stark effect should probably give correct results, because in other cases the method had worked so well. On the other hand, there is very little difference between a constant electric field and a slowly varying electric field. When we have an electric field varying, not very slowly, but say with a frequency which comes near to the orbital frequency, then we know of course that resonance takes place, not when the frequency of the outer electric field coincides with the frequency of the orbit, but when it coincides with the frequency given by the transitions, and observed in the spectrum.

When we discussed this problem at length, Bohr tried to say that as soon as the electric field varies with time, then forces of radiation come in, and therefore it may not be possible to calculate the result in a classical way. But of course at the same time he saw that it is rather artificial to invoke the forces of radiation at this point. We were therefore soon inclined to say that there must be something wrong with the mechanical model of the discrete stationary state. There was one very decisive paper which has not been mentioned yet. It was a paper of Pauli on the H_2^+ ion. Pauli thought that we could possibly apply the Bohr-Sommerfeld quantization rules when we had a well-defined model with periodic orbits, as in hydrogen; but perhaps not in a model so complicated as the helium atom, say, where two electrons move around the nucleus; because there we would get into all the terrible mathematical difficulties and complications of the three body problem. On the other hand, when we have two fixed centers, two hydrogen nuclei and one electron, then the motion of the electron is still a nice periodical motion and can be calculated. For the rest, the model is already rather complicated; hence it can be used as a check on whether the old rules really apply in such an intermediate case. Pauli did work out this model and found that he actually did not get the correct energy of the H_2^+ by his calculations. So the doubts about the use of classical mechanics for the calculation of discrete stationary states increased, and attention was shifted more and more to the transitions between the stationary states. We understood that in order to get the whole explanation of the phenomena it was not sufficient to calculate the energy. One also had to calculate transition probabilities. We knew from Einstein's paper of 1918 that the transition probabilities are defined as quantities referring to two states, initial and final state. Bohr had pointed out in his correspondence principle that these transition probabilities could be estimated by connecting them with the intensities of higher harmonies in the Fourier expansion of the electronic orbit. The idea was that every line corresponds to one Fourier component in the expansion of the electronic motion; from the square of this amplitude one can calculate the intensity. This intensity, of course, is not immediately connected with Einstein's transition probability, but it is related to it, so it allows some estimate of Einstein's quantities. In this way attention gradually moved over from the energy of the stationary state to the transition probability between stationary states, and it was Kramers who started seriously to study the dispersion of an atom, and to relate the behavior of Bohr's model under radiation with the Einstein coefficients.

In writing down a dispersion formula, Kramers was guided by the idea of virtual harmonic oscillators in the atom, corresponding to the harmonics. He and I also discussed scattering phenomena, where the frequency of the scattered light is different from the frequency of the incident light. Here the scattered light-quantum is different from the incoming quantum, because during the scattering the atom makes a transition from one state to another. Such phenomena had just been discovered by Raman in band spectra. When one tried to write down formulae for the dispersion in these cases, one was forced to speak not only about the transition probabilities of Einstein, but also about transition amplitudes; one had to give phases to these amplitudes, and to multiply two amplitudes—the amplitude going from state m to state n, say, with the amplitude going from n to state k or whatever, and then to sum over the intermediate states n; only on doing that did we get reasonable formulae for the dispersion.

So you see that by fixing attention not on the energy of the stationary state, but on the transition probabilities and on dispersion, one eventually came into a new way of looking at things; actually, as I just said, these sums of products, which Kramers and I had written into our paper on dispersion, were already almost products of matrices. So it was only a very small step from there to saying: Well, let us abandon this whole idea of the electronic orbit, and let us simply replace the Fourier components of the electronic orbit by the corresponding matrix elements. At that time I must confess I did not know what a matrix was, and did not know the rules of matrix multiplication. But one could learn these operations from physics, and later on it turned out that it was matrix multiplication, well known to the mathematicians.

By this time you see that the idea of an electronic orbit, connected with the discrete stationary state, had been practically abandoned. The concept of the discrete stationary state had, however, survived. This concept was necessary, and had its basis in the observations. But the electronic orbit could not be connected with observations and had therefore been abandoned, and what had remained were these matrices for the coordinates.

I should perhaps mention that already, before this had happened in 1925, Born, in his Göttingen seminar in 1924, had emphasized that it was wrong to put the blame for the difficulties of quantum theory solely on the interaction between radiation and the mechanical system. He propagated the idea that mechanics had to be revised and replaced by some kind of quantum mechanics in order to supply the basis for an understanding of atomic phenomena. And then matrix multiplication was defined. Born and Jordan, and independently Dirac, discovered that those extra conditions, which had been added to matrix multiplication in my first paper, can actually be written in the elegant form

$$pq - qp = \frac{h}{2\pi i}.$$

Thereby they were able to establish a simple mathematical scheme for quantum mechanics.

But even then one could not say what this discrete stationary state really was, and therefore I now come to the second part of my talk—the concept of a "state." In 1925 one did have a method for calculating the discrete energy values of the atom. One also had, at least in principle, a method for calculating the transition probabilities. But what was this state of the atom? How could it be described? It could not be described by referring to an electronic orbit. So far, it could be described only by stating an energy and transition probabilities; but there was no picture of the atom. Furthermore, it was clear that sometimes there are non-stationary states. The simplest example of a non-stationary state was an electron moving through a cloud chamber. So the question really was, how to handle such a state, which can occur in nature. Can such a phenomenon as the path of the electron through a cloud chamber be described in the abstract language of matrix mechanics?

Fortunately, at that time, wave mechanics had been developed by Schrödinger. And in wave mechanics things looked very different. There one could define a wave function for the discrete stationary state. For some time Schrödinger thought that the following picture of a discrete stationary state could be developed. One had a three-dimensional standing wave, which can be written as the product of a function in space and a periodical e^{iwt} of time, and the absolute square of this wave function meant the electric density. The frequency of this standing wave was to be identified with the term in the spectral law. This was the decisive new point in Schrödinger's idea. These terms did not necessarily mean energies; they just meant frequencies. And so Schrödinger arrived at a new "classical" picture of the discrete stationary state, which at first he believed could actually be applied in atomic theory. But then it soon turned out that even that was not possible. There were very heated discussions in Copenhagen in the summer of 1926. Schrödinger thought that the wave picture of the atom—with continuous matter spread out around the nucleus, according to its wave function—could replace the older models of quantum theory. But the discussions with Bohr led to the conclusion that this picture could not even explain Planck's law. It was extremely important for the interpretation to say that the eigenvalues of the Schrödinger equation are not only frequencies—they are actually energies.

In this way, of course, one came back to the idea of quantum jumps from one stationary state to the other, and Schrödinger was very dissatisfied with this result of our discussions. But even when we knew this, and accepted the quantum jumps, we did not know what the word *state* could mean. One could, of course, try—and it was tried very soon—to see whether one could describe the path of the electron through a cloud chamber by means of Schrödinger's wave mechanics. It turned out that this was not possible. In its initial position, the electron could be represented by a wave packet. This wave packet would move along and thereby one got something like the path of the electron through the cloud chamber. But the difficulty was that this wave packet would become bigger and bigger, so that, if only the electron ran long enough, it might have a diameter of one centimeter or more. This is certainly not what we see in the experiments, and so this picture again had to be abandoned. In this situation, of course, we had many discussions, difficult discussions, because we all felt that the mathematical scheme of quantum or wave mechanics was already final. It could not be changed, and we would have to do all our calculations from this scheme. On the other hand, nobody knew how to represent in this scheme such a simple case as the path of an electron through a cloud chamber. Born had made a first step by calculating from Schrödinger's theory the probability for collision processes; he had introduced the notion that the square of the wave function was not a charge density, as Schrödinger had believed; that it meant the probability of finding the electron at a given place.

Then came the transformation theory of Dirac and Jordan. In this scheme, one could transform from $\psi(\vartheta)$ to, for instance, $\psi(\rho)$, and it was natural to assume that the square, $|\psi(\rho)|^2$, would be the probability of finding the electron with momentum ρ. So gradually one acquired the notion that the square of the wave function, which, by the way, was not a wave function in three-dimensional space, but in configuration space, meant the probability of something. With this knowledge, we returned to the electron in the cloud chamber. Could it be that we had asked the wrong question? I remembered Einstein telling me, "It is always the theory which decides what can be observed." And that meant, if it was taken seriously, that we should not ask, "How can we represent the path of the electron in the cloud chamber?" We should ask instead, "Is it not perhaps true that, in nature, only such situations occur as can be represented in quantum mechanics or wave mechanics?"

Turning the question around, one saw at once that this path of an electron in a cloud chamber was not an infinitely thin line with well-defined positions and velocities; actually, the path in the cloud chamber was a sequence of points which were not too well-defined by the water droplets, and the velocities were not too well-defined either. So I simply asked the question, "Well, if we want to know of a wave packet both its velocity and its position, what is the best accuracy we can obtain, starting from the principle that only such situations are found in nature as can be represented in the mathematical scheme of quantum mechanics?" That was a simple mathematical task and the result was the principle of uncertainty, which seemed to be compatible with the experimental situation. So finally one knew how to represent such a phenomenon as the path of an electron, but again at a very high price. For this interpretation meant that the wave packet representing the electron is changed at every point of observation, that is, at every water droplet in the cloud chamber. At every point we get new information about the state of the electron; therefore we have to replace the original wave packet by a new one, representing this new information.

The state of the electron, thus represented, does not allow us to ascribe to the electron in its orbit definite properties such as coordinates, momentum and so on. All we can do is to speak about the probability of finding—under suitable experimental conditions—the electron at a certain point, or a certain value for its velocity. So finally we have come to a definition of state which is much more abstract than the original electronic orbit. Mathematically, we describe it by a vector in Hilbert space, and this vector determines probabilities for the results of any kind of experiment which can be carried out at this state. The state may change with every new addition to our information.

This definition of state made a very big change in the description of natural phenomena, and I doubt whether the unwillingness of Einstein, Planck, von Laue and Schrödinger to accept it should be attributed simply to prejudice. The word *prejudice* is too negative in this context, and does not cover the situation. It is of course true that Einstein, for instance, thought it must necessarily be possible to give some kind of objective description of the state of affairs, the state of an atom, in the same sense as had been possible in the older physics. But it was indeed extremely difficult to give up this notion, because all our language is bound up with this concept of objectivity. So all the words which we use in physics in describing experiments, such as the words *measurement* or *position* or *energy* or *temperature* and so on, are based on classical physics and its idea of objectivity. The statement that such an objective description is not possible in the world of the atoms, that we can only define a state by a direction in Hilbert space—such a statement was indeed very revolutionary; and I think it is really not so strange that many physicists of that time were simply not willing to accept it.

I had a discussion with Einstein about this problem a few months before his death in 1954. It was a very nice afternoon that I spent with Einstein, but still, when it came to the interpretation of quantum mechanics, I could not convince him and he could not convince me. He always said, "Well, I agree that any experiment, the results of which can be calculated by means of quantum mechanics, will come out as you say; but still, such a scheme cannot be a final description of nature."

We now come to the third concept I wanted to discuss, the concept of the elementary particle. Before the year of 1928, every physicist knew what we meant by an elementary particle. The electron and the proton were the obvious examples, and at that time we would have liked simply to take them as point charges, infinitely small, defined simply by their charge and their mass. We had to agree reluctantly that they must have a radius, since their electromagnetic energy had to be finite. We did not like the idea that such objects should have properties like a radius, but still we were happy that at least they seemed to be completely symmetrical, like a sphere. But then the discovery of electronic spin changed this picture considerably. The electron was not symmetrical. It had an axis, and this result emphasized that perhaps such particles have more than one property, and that they are not simple, not so elementary as we had thought before. The situation was again changed completely in 1928, when Dirac developed the relativistic theory of the electron and discovered the positron. A new idea cannot be quite clear from the beginning. Dirac thought at first that the negative energy holes of his theory could be identified with the protons; but later it became clear that they ought to have the same mass as the electron, and finally they were discovered in experiments, and were called positrons. I think that this discovery of antimatter was perhaps the biggest change of all the big changes in physics of our century. It was a discovery of utmost importance, because it changed our whole picture of matter. I would like to explain this in more detail in the last part of my talk.

First, Dirac suggested that such particles can be created by the process of pair production. A light quantum can lift a virtual electron from one of these negative energy states in the vacuum to a higher positive energy, and that means that the light quantum has created a pair—of electron and positron. But this meant at once that the number of particles was not a good quantum number anymore, there was no conservation law for the number of particles. For instance, according to Dirac's new idea, one could say that the hydrogen atom does not necessarily consist of proton and electron. It may also temporarily consist of one proton, two electrons, and one positron. And actually, when one takes the finer details of quantum electrodynamics into account, these possibilities do play some role.

In any case of interaction between radiation and electron, such phenomena as pair production can happen. But then it was natural to assume that similar processes may occur in a much wider range of physics. We had known since 1932 that there are no electrons in the nucleus, that the nucleus consists of protons and neutrons. But then Pauli suggested that beta decay could be described by saying that an electron and a neutrino are being created in beta decay. This possibility was formulated by Fermi, in his theory of beta decay. So you see that already at that time the law of conservation of particle number was completely abandoned. It was understood that there are processes in which particles are created out of energy. The possibility of such processes was, of course, already allowed by the theory of special relativity, when energy is transmuted into matter. But its reality occurred for the first time in connection with Dirac's discovery of antimatter and pair-creation.

The theory of beta decay, if I recall correctly, was published by Fermi in 1934. A few years later in connection with cosmic radiation, we asked the question: What happens if two elementary particles collide with very high energy? The natural answer was that there was no good reason why one should not have many particles created in such an act. So actually, the hypothesis of multiple production of particles in high energy collisions was a very natural assumption, after Dirac's discovery. It was checked experimentally only fifteen years later, when one studied very high energy phenomena and could observe such processes in the big machines. But when it was known that, in very high energy collisions, any number of particles can be created, provided only that the initial symmetry is identical with the final symmetry, then it also had to be assumed that any particle is really a complicated compound system, because with some degree of truth one can say that any particle consists of virtually any number of other particles. Of course, we would still agree that it may be a reasonable approximation to consider a pion as composed only of nucleon and antinucleon, and that we should not consider higher compositions. But that is only an approximation, and if we have to speak rigorously, then we should say that in any one pion we have a number of configurations of several particles, up to an arbitrarily high number of particles, if only the total symmetry is the same as the symmetry of the pion. So it was one of the most spectacular consequences of Dirac's discovery, that the old concept of the elementary particle collapsed completely. The elementary particle was not elementary anymore. It is actually a compound system, rather a complicated many-body system, and it has all the complications which a molecule or any other such object really has.

There was another consequence of Dirac's theory which is important. In the old theory, let us say in non-relativistic quantum theory, the ground state was an extremely simple state. It was just the vacuum, the empty void and nothing else, and had therefore the highest possible symmetry. In Dirac's theory, the ground state was different. It was an object which was filled with particles of negative energy that could not be seen. Besides that, if the process of pair-production is introduced, one should expect that the ground state must contain a probably infinite number of virtual pairs of positrons and electrons, or of particles and antiparticles; so you see at once that the ground state is a complicated dynamical system. It is one of the eigensolutions defined by the underlying natural law. If the ground state is to be interpreted in this way, one can further see that it need not be symmetrical under the groups of the underlying natural law. In fact, the most natural explanation of electrodynamics seems to be that the underlying natural law is completely invariant under the isospin group, while the ground state is not. The assumption that accordingly the ground state is degenerate under rotations in isospace enforces the existence of long-range forces, or of particles with rest mass zero, following a theorem of Goldstone. Coulomb interaction and photons should probably be interpreted in this way.

Finally, Dirac—in consequence of his theory of holes—has propagated the idea, in his Bakerian lecture of 1941, that in a relativistic field theory with interaction, use should be made of a Hilbert space with indefinite metric. It is still a controversial question, whether this extension of conventional quantum theory is really necessary. But, after many discussions during the last decades, one cannot doubt that theories with indefinite metric can consistently be constructed, and can lead to a reasonable physical interpretation.

So the final result at this point seems to be that Dirac's theory of the electron has changed the whole picture of atomic physics. After abandoning the old concept of the elementary particles, those objects which had been called "elementary particles" have now to be considered as complicated compound systems, and will have to be calculated some day from the underlying natural law, in the same way as the stationary states of complicated molecules will have to be calculated from quantum or wave mechanics. We have learned that energy becomes matter when it takes the form of elementary particles. The states called elementary particles are just as complicated as the states of atoms and molecules. Or, to formulate it paradoxically: every particle consists of all other particles. Therefore we cannot hope that elementary particle physics will ever be simpler than quantum chemistry. This is an important point, because even now many physicists hope that some day we might discover a very simple route to elementary particle physics, as the hydrogen spectrum was in the old days. This, I think, is not possible.

In conclusion I would like again to say a few words about what have been called "prejudices." You may say that our belief in elementary particles was a prejudice. But again I think that would be too negative a statement, because all the language which we have used in atomic physics in the last 200 years is based, directly or indirectly, on the concept of the elementary particle. We have always asked the question: "Of what does this object consist, and what is the geometrical or dynamical configuration of the smaller particles in the bigger object?" Actually, we have always gone back to this philosophy of Democritus; but I think we have now learned from Dirac that this was the wrong question. Still, it is very difficult to avoid questions which are already part of our language. Therefore it is natural that even nowadays many experimental physicists, and perhaps even some theoreticians, still look for really elementary particles. They hope, for instance, that quarks, if they exist, may be able to play this role.

I think that this is an error. It is an error because, even if the quarks were to exist, we could not say that the proton consists of three quarks. We would have to say that it may temporarily consist of three quarks, but may also temporarily consist of four quarks and one antiquark, or five quarks and two antiquarks and so on. And all these configurations would be contained in the proton, and again one quark might be composed of two quarks and one antiquark and so on. So we cannot avoid this fundamental situation; but since we still have the questions from the old concepts, it is extremely difficult to stay away from them. Very many physicists have looked for quarks, and will probably do so in the future. There has been a very strong prejudice in favor of quarks during the last ten years, so I think they ought to have been found, if they existed. But that is a matter to be decided by the experimental physicists.

There remains the question: "What then has to replace the concept of a fundamental particle?" I think we have to replace this concept by the concept of a fundamental symmetry. The fundamental symmetries define the underlying law which determines the spectrum of elementary particles. I shall not now go into a detailed discussion of these symmetries. From a careful analysis of the observations I would conclude that, besides the Lorentz group, also SU_2, the scaling law, and the discrete transformations P,C,T are genuine symmetries; but I would not include SU_3, or higher symmetries of this type, among the fundamental symmetries; they may be produced by the dynamics of the system as approximate symmetries.

But this is again a matter which will have to be decided by the experiments. I only wanted to say that what we have to look for are not fundamental particles, but fundamental symmetries. And when we have actually made this decisive change in the concepts, which came about by Dirac's discovery of antimatter, then I do not think we need any further breakthroughs to understand the elementary—or rather non-elementary—particles. We must only learn to work with his new and unfortunately rather abstract concept of the fundamental symmetries; but this may come, in time.

The Beginnings of Quantum Mechanics in Göttingen*

Quantum mechanics originated in Göttingen fifty years ago, and this jubilee provides a good occasion to report here on the beginnings of this development, in the tradition of the old colloquium. In so doing, I cannot and do not wish to take on the role of the historian, who attempts a thorough study of sources and the delineation of a maximally correct, objective picture of individual events; there are very good historical accounts and I would not be able to improve on them. I should like, rather, to paint a subjective picture, to depict details that are not in the history books, and to say what steps appeared most important to me, even though their importance must perhaps be judged differently from the objective point of view. But before I begin on this, a word should doubtless be said about the geographical situation of Göttingen in the landscape of the physics, and especially the atomic physics, of the period in question. Planck's quantum theory, in those days, was really not a

---

*Unpublished manuscript of a lecture, intended for delivery in Göttingen on May 26, 1975.

theory, but an embarrassment. Into the well-founded edifice of classical physics it brought ideas that led, on many points, to difficulties and contradictions, and hence there were not many universities where there was any desire to tackle these problems seriously. Other than in Copenhagen, Bohr's theory was primarily taught and developed by Sommerfeld in Munich, and it was only in 1920, with the appointment of Franck and Born, that the Göttingen faculty finally decided to join this scientific movement. If we compare the three centers, Copenhagen, Munich and Göttingen, where the subsequent development primarily took place, we can relate them to three lines of work in theoretical physics which are still clearly distinguishable even today: the phenomenological school attempts to unite new observational findings in an intelligible fashion, to present their connection by means of mathematical formulae which appear to some degree plausible from the standpoint of current physics; the mathematical school endeavors to represent natural processes by means of a carefully worked-out mathematical formalism, which also satisfies to some extent the mathematicians' demands for rigor; the third school, which may be called conceptual or philosophical, tries above all to clarify the concepts by means of which events in nature are ultimately to be described. In retrospect we may align Sommerfeld's school in Munich with the phenomenological approach, the Göttingen center with the mathematical, and the Copenhagen group with the philosophical tendency, though the transitions are naturally fluid. My report will divide into three periods, the preparatory years of 1922–1924, the decisive year 1925, and the years of exploitation, 1926 and 1927.

In speaking of the beginnings of quantum mechanics in Göttingen, we must undoubtedly start with the Bohr Festival in 1922. At the instigation of Hilbert and the physicists Franck, Born and Pohl, the university had asked the Dane, Niels Bohr, to give a comprehensive series of lectures on his theory. Guests were invited from outside, including Sommerfeld from Munich, and the whole gathering, as one of the first after the great economic stringency of the post-war period, bore the marks of a joyous new beginning; a new beginning in the international relations of science, but also in the tasks of the newborn atomic physics. Not only this, but Göttingen, in gorgeous summer weather, was resplendent with gardens and flowers, and so, despite the difficulty of the subject, the atmosphere, and the excitement of the students who filled most of the auditorium, gave the lectures such a festive air, that the name Bohr Festival soon made the rounds—in allusion to the Handel Festivals in the Göttingen civic theater, which had just then begun. Sommerfeld had brought me along from Munich; he had generously financed my trip, which would have been far beyond my resources at that time, and I enjoyed the festive days to the full, though often on an empty stomach, as was normal in those days for a student in the fourth semester.

Bohr set out his theory in full detail in the lectures, and from the outset there were two issues, above all, which riveted my interest, and doubtless also that of many other listeners: firstly the question, whether the energies of discrete stationary states could really be determined with universal correctness by applying the Bohr-Sommerfeld quantum conditions to the mechanical motions of electrons in the atom; and secondly the question, how well Bohr's many-electron atomic models fitted in with the chemical and optical data relating to the periodic system of the elements.

As to the first issue, I thought that I soon detected from Bohr's utterances that he believed less firmly than did Sommerfeld, say, in the applicability of classical mechanics to the motions of electrons within the atom. The fact that, on this assumption, the orbital frequencies of the electrons could not coincide with the frequencies of the radiation emitted by the atom, was felt even by Bohr himself to be an almost intolerable contradiction, which he tried merely to patch over in desperation with the idea of his correspondence principle. A discussion question that I put to Bohr in this connection then led to a long conversation, on a walk over the Hainberg, from which I learnt for the first time how difficult, indeed how well-nigh hopeless, these problems of atom dynamics then appeared. Bohr stressed repeatedly that human language is obviously inadequate to describe processes within the atom, since we are dealing there with a realm of experience totally shut off from direct inspection. But since all understanding and communication among physicists depends on language, we can meanwhile think of no solution at all. At that time Bohr still believed, indeed, that the difficulties were to be looked for, initially, in the theory of radiation, and hence in electrodynamics, whereas I, on the contrary, felt I had gained from the discussion increasing indications that mechanics, and perhaps even kinematics, was the scapegoat. On the second issue, the quantization of many-electron systems and the periodic table of the elements, we said but little at that time. Bohr confirmed to me, what Pauli and I in Munich had long suspected, that he had not worked out the complex atomic models by classical mechanics; they had come to him intuitively, rather, on the basis of experience, as pictures—so far as mere mechanical pictures can be suitable at all—for representing events within the atom.

Bohr's lectures provided a crucial stimulus to the further development of atomic physics in Göttingen. Since I was studying there during the winter semester of 1922–23— Sommerfeld had gone to America during this period—I was able to follow out this influence from the start. Born instituted a seminar on the problems of Bohr's theory. Since, as I recollect, only some eight physicists and mathematicians took part, the seminar was often held in the evening at Born's house. Frau Born sustained it with cakes or fruit. I can no longer give a complete list of the participants; Jor-

BEGINNINGS OF QUANTUM MECHANICS / 41

dan, Hund, Fermi, Pauli, Northeim and the mathematician Karekjarto certainly attended, perhaps not always at the same time. But here, too, I would sooner leave it to the historians to recover the details exactly. The tasks assigned to us by Born in the context of this seminar all came from the field of mechanics, and from this it was already evident that Born, too, was looking for the real difficulties in mechanics, and not primarily in electrodynamics or the theory of radiation. In this connection I was given the job of busying myself with the classical perturbation theory of the astronomers; for by then it was already clear to all concerned that it was not enough to deal with the simple one-electron problem of hydrogen. For hydrogen, even when it is affected by external electromagnetic fields, the Bohr-Sommerfeld rules were exceedingly successful; but for systems having several electrons, insuperable difficulties arose. From the Göttingen mathematicians one could learn in particular of the notorious difficulties of the astronomical three-body problem. Periodic and non-periodic solutions there lie densely close together. The quantum conditions, on the other hand, certainly rested on the assumption of periodic solutions. So we began by delving into the general perturbation theory of Hamilton-Jacobi mechanics, as utilized by the astronomers. Later we studied resonance effects between different planetary orbits in the same system; on one occasion I had to report on the so-called Bolin method. The real value of these efforts was the realization that, although the classicial mechanics could not be correct, it did, however, contain many features that are also re-encountered in the empirical regularities of quantum theory, and that Bohr's correspondence principle somehow formed the bridge between these two otherwise so different conceptions. So whereas in Munich the exact calculation of individual states was regarded as the most important success of the quantum theory, while the correspondence principle seemed, rather, to be a less happy stop-gap, in the Göttingen discussions the correspondence principle came to occupy an increasingly central position. This was supplemented by the fact that in the phenomenological studies of the Munich school on the anomalous Zeeman effect, and on line-intervals and intensities in multiplets, formulae had repeatedly emerged which looked very like those that could be derived from classical mechanics. In such formulae, for example, the square of the angular momentum often appeared; but if the quantum number of the angular momentum of the system was I, then empirically the square was not I^2, but $I(I+1)$, so that in a study of the Zeeman effect I designated the root of this expression as the angular momentum. This led Sommerfeld, who held integers to be crucial, to say that the quantity I had introduced as I was nebulous. Thus very gradually the feeling grew up in the Göttingen discussions, that the classical formulae were indeed half correct, but still only half correct, and that with some ingenuity the correct quantum-theoretical formulae could be conjectured from them.

Here in Göttingen, also, the other promptings from the Bohr Festival, concerning many-electron systems and the periodic table, were, of course, further pursued. I seem to remember that this took place primarily in discussions between Born and Hund, while I, despite the anomalous Zeeman effect and the multiplets, which still preoccupied me from the Munich period, was applying myself more to the essential problems of the correspondence principle. An important further stimulus in this direction came from studies on the dispersion theory of Ladenburg and Kramers. Here the Fourier components of classical orbital motion were brought into relation with the Einsteinian transition-probabilities in the dispersion of light. Bohr's correspondence principle was thus interpreted in detail by way of relationships from the classical theory of dispersion, so that again the classical mechanics could be recognized as half correct.

The state of the discussions at that time is very accurately presented in a study by Born, dating from the summer of 1924, which bears the title, *On Quantum Mechanics*. Here, therefore, the term "quantum mechanics" appears for the first time, and perhaps I ought to read out the summary at the beginning of the work. It runs: "The paper contains an attempt to establish the first step towards a quantum mechanics of coupling, which provides an account of the most important properties of atoms, stability, resonance for transition frequencies, the principle of correspondence, and arises naturally from the classical laws. This theory contains the dispersion formulae of Kramers, and therefore shows close affinity with the Munich formulations of the rules governing the anomalous Zeeman effect." As emerges from the detail of the work, Born was very clearly of the feeling that quantum mechanics differs from classical mechanics, in that the differential equations of the classical theory have to be replaced in quantum theory by difference equations. He therefore gave me the task of studying the theory of difference equations, which had already, of course, been extensively developed by the mathematicians. I did this with considerable aesthetic satisfaction, but still with the feeling that physical problems could never be resolved by pure mathematics. The real obstacle, which we suspected, indeed, at that time, but did not yet understand, was the fact that we were still talking of electron pathways, and were really compelled to do so; for electron tracks were certainly visible in the cloud-chamber, so there also had to be electron pathways in the interior of the atom.

Before now arriving at the developments of 1925, I would like just to tell of two small episodes, which show how intensively we were then preoccupied with the problems of quantum theory. The group of young people studying with Born and Franck was really no longer able to talk of anything but quantum theory, so full we were of its successes and inner contradictions. We used at that time to take our modest midday meal at a private dining-place across from the college building. One day, to my astonishment, I was summoned by our hostess after the meal to a privy conference in her room. She explained to me that, alas, we physicists could no longer eat there in future, since the everlasting shop-talk at our table was so unbearable to those at the other tables, that she would lose her remaining customers if she continued to entertain us. On another occasion we had gone off together on a ski trip to the Harz, probably to climb the Brocken, and on the way back to Andreasberg one of the group, I think it was Hanle, had got lost. We searched and could not find him, and were beginning to fear that he might be hurt, or have somehow missed his way in the forest. Suddenly we heard from a quite distant patch of woods a rather plaintive voice crying $\eta\nu$, and then we knew where we had to look.

But now to the events of 1925. In the winter semester of 1924–25 I had again been working in Copenhagen, developing the dispersion theory along with Kramers. In this connection, certain mathematical expressions had appeared in the formulae for the Raman effect, which in classical theory figured as products of Fourier series, whereas in quantum theory they obviously had to be replaced by similarly constructed products of series having to do with the quantum-theoretical amplitudes for emission or absorption lines. The multiplication law for these series looked simple and convincing. When, in the summer semester of 1925, I again took up the work in Göttingen, one of the earliest discussions with Born led to the conclusion that I should try to conjecture the correct amplitudes and intensities for hydrogen from the correspondence-type matching formulae of the classical theory. This method of conjecture had proved itself. We thought we had learned it sufficiently from earlier work. But on going into it more deeply the problem turned out to be too complex, at least for my mathematical capacities, and I searched for simpler mechanical systems, in which the method of conjecture promised more success. In so doing I had the feeling that I should renounce any description of electron pathways, indeed that I ought deliberately to repress such ideas. I wanted, rather, to trust entirely to the half-empirical rules for the multiplication of amplitude series, which had proved themselves in dispersion theories. I chose as a mechanical system the one-dimensional anharmonic oscillator, which seemed to me simple enough, and yet not too trivial.

About this time, at the end of May or beginning of June, I had to ask Born for two weeks leave, since I was suffering from a really unpleasant bout of hay-fever, and wanted to recover on the lonely island of Heligoland, far from all flowering fields. There I was then able to devote myself to my problem, without any external interruption. I replaced the positional co-ordinates, therefore, with a table of amplitudes that was meant to correspond to the classical Fourier series, and wrote out the classical equation of motion for it, making use as I did so, in the non-linear term representing the anharmonicity, of the multiplication of amplitude-series that had proved itself in dispersion theory. Only much later did I learn from Born that it was simply a matter here of multiplying matrices, a branch of mathematics that had hitherto remained unknown to me. It bothered me, that in this way of multiplying series, $a \times b$ was not necessarily equal to $b \times a$. By means of the equation of motion, however, the tables representing the position were not yet unequivocally determined. A substitute for the Bohr-Sommerfeld quantum condition still had to be found; for the latter, of course, employed the concept of electron pathways, which I had expressly forbidden myself. But a correspondence-type transformation soon led to the addition rule, derived by Thomas and Kuhn from dispersion theory, and known to me from my time in Copenhagen. With this, it seemed, the whole mathematical schema was established, and it was now necessary to inquire whether it could be interpreted as mechanics. For this purpose it had to be shown that there is an expression for the energy, that can be represented by the tables of coordinates and matches by correspondence the classical formula for energy; that this expression is constant over time, *i.e.*, that the energy principle holds; and hence that the energy table is, as we now call it, a diagonal matrix. Finally, it had to be demonstrated, that the differences of the energy values in different states yield, up to the factor h, namely Planck's constant, the frequency of the radiation emitted in the transition. That was a large number of conditions that had to be satisfied; the calculations were elementary, but also, on that account, extremely detailed. In the end it turned out that all conditions were met, and hence that one could hope to have found the basis for a quantum mechanics. After my return to Göttingen, I showed the paper to Born, who found it interesting but somewhat disconcerting, inasmuch as the concept of electron pathways was totally eliminated. But he sent it for publication to the *Zeitschrift für Physik*. Born and Jordan now plunged into the mathematical consequences of the paper, this time in my absence, since I had been invited by Ehrenfest and Fowler to give lectures in Holland and Cambridge, England. In a few days Born and Jordan found the key relation

$$pq - qp = \frac{h}{2\pi i}$$

by means of which the whole mathematical schema could be made clear; in particular, such principles as that of the conservation of energy could now be derived with ease and elegance.

When I first of all returned to Copenhagen in September, Born and Jordan, as I recollect, had already completed their work, which contains a convincing mathematical foundation for quantum mechanics. Some time later—it must have been the end of October—when I was back in Göttingen, I received a letter from Dirac in Cambridge, in which he communicated to me the form of quantum mechanics that he had drawn up on the basis of my reports in Cambridge. He made no use of the matrix calculus, but introduced for the dynamical variables p and q an algebra in which, of course, the commutation relation played the decisive role. It could be seen at once that Dirac's formulation was equivalent to the method of Born and Jordan. We concluded, therefore, that in the new mechanics we were standing on relatively firm mathematical ground, and decided *à trois*—that is, Born, Jordan and I—to write a comprehensive paper dealing with systems of many degrees of freedom, the quantum-mechanical theory of perturbation, and the relationships to radiation theory. In this work, the mathematical tradition of Göttingen University stood us in good stead. Born was not only fully acquainted with the mathematicians' theory of matrices; he also knew Hilbert's theory of integral equations and of the quadratic forms of an infinite number of variables. He could therefore prove that treatment of a quantum-mechanical system amounted to a principal-axis transformation of infinite quadratic forms. From this the theory of perturbations could also be easily derived. In Jordan's calculations on oscillatory phenomena, the discontinuous character of the quantum jumps became plainly visible.

But there were also difficulties in the writing, which have been mentioned by Born in his memoirs. It was of prime importance to me to give pre-eminence to the physical content of the theory, especially the absence of electron pathways in the atom, whereas Born considered the principal-axis transformation—a mathematical formalism—to be the heart of the theory. There was also an external difficulty, in that Born was going to America at the end of October; thus we only had a few days of collective discussion in Göttingen, and Jordan and I had to finish off the paper after Born had already departed. You can see how, even in those days, scientific progress was hampered by foreign travel, lectures and conferences, although not yet to the frightful extent it is nowadays. The three-man paper, as we then called it, since team-work was still not the rule in those days, was sent off in mid-November to the *Zeitschrift für Physik*.

We were at that time in regular correspondence with Pauli in Hamburg, he having belonged to the inner circle of quantum theorists from the beginning, in Munich and Göttingen. Pauli undertook to apply the new quantum mechanics to the problem of hydrogen, which had played so large a part in the history of the quantum theory. Using a method derived from Lenz in Hamburg, he was entirely successful, and already before the three-man paper was completed in Göttingen, had shown that the new theory also gave the correct spectrum for the hydrogen atom. Pauli was also able to deal exactly with the more complex case of the hydrogen atom in crossed electric and magnetic fields. With this success, the persuasiveness of the new mechanics was decisively enhanced. I have now sketched the most important advances in quantum theory during the year 1925, so far as they were directly connected with Göttingen, and I ought perhaps to add a few words about the difficulties previously referred to, concerning the problem of "physical content *versus* mathematical form."

I was, of course, entirely clear at that time about the enormous significance of the self-contained and beautiful mathematical form that Born and Jordan had given to the new theory. In no other city in the world could this mathematical picture have been so rapidly evolved as in Göttingen. But from the start I still had the feeling that the greatest difficulties did not lie in the mathematics, but at the point where the mathematics had to be linked to nature. In the end, after all, we wanted to describe nature, and not just do mathematics, and I felt that even the three-man paper had still done nothing to resolve this problem. We could, indeed, calculate the energy of stationary states, or the intensity of lines, but how we were to describe the path of an electron in the cloud-chamber, which can actually be seen directly, we did not know. We had resolved not to speak of paths, but in the end the paths still obviously belonged in some way to physical reality. After completing the three-man paper, I wrote an unhappy letter to Pauli, to whom, indeed, I always confided my worries, and I should like to read a passage from this letter: "I have taken every trouble to make the paper more physical than it was, and am half content with it thus. But I am still pretty unhappy about the whole theory, and was so pleased that in your view about mathematics and physics you are so entirely on my side. Here I am in an environment which thinks and feels in exactly the opposite way, and I don't know whether I am just too stupid to understand the mathematics. Göttingen falls into two camps: one, which like Hilbert or even Weyl, who, in a letter to Jordan, speaks of the great success achieved in physics by introduction of the matrix calculus; the other, which says like Franck that we still can never understand the matrices." In reality there had doubtless been a collision at this point between the two modes of working in theoretical physics, which I distinguished at the outset as the conceptual and the mathematical, and assigned to the two cities of Copenhagen and Göttingen. The mathematical formulation was still not yet adequate to the conceptual formulation. This can also be seen very clearly from the history of relativity theory. Lorentz, with his transformation formulae, had in essence discovered the mathematical formulation, but it was only Einstein who furnished conceptual enlightenment. Lorentz, too, had doubtless had an inkling of the conceptual problem, since he brought in an apparent time alongside the absolute time of earlier physics; but the whole business was not really understood until some years later, thanks to Einstein.

The state of our inquiries in Göttingen toward the end of 1925 could thus be summarized more or less as follows: the mathematical formalism of quantum mechanics stood firm, but, as it later turned out, was not yet completely developed. As to how the formalism was to be applied to experience, there were indeed some definite things to be said, but the real conceptual enlightenment had not yet been attained.

The year 1926 began with a surprise. At first by hearsay, and then in the form of proof-sheets, we got news of Schrödinger's first paper on wave-mechanics, in which the determination of energy values in the hydrogen atom was reduced to an eigen-value problem for three-dimensional matter-waves. The physical picture that Schrödinger started from, and which derived from de Broglie, looked altogether unlike the Bohr model of the atom with which we had been working hitherto. But the results were the same, and there were important formal similarities. The concept of an electron pathway was lacking in Schrödinger, just as in the Göttingen quantum mechanics, and in both theories the determination of energy values for stationary states amounted to an eigen-value problem in linear algebra. The suspicion that the two theories were mathematically equivalent, that is, transformable into each other, arose very quickly and was generally discussed, as it also was in our correspondence with Pauli. Already, by the end of 1925, Born in America had produced, in collaboration with Norbert Wiener, a new mathematical formulation of quantum mechanics, which employed the concept of a linear operator, and which, as is ascertainable by hindsight, could easily have led to Schrödinger's formalism of wave-mechanics. In fact, however, Born and Wiener did not find this transition. By March 19, 1926, Schrödinger was able to send the desired proof of equivalence to the *Annalen der Physik*. But Pauli, too, as I recollect, already had this proof quite early on, and communicated it to me in a letter, though he did not publish it at the time. Here, too, I am not quite certain in my memory. Thus we knew, at all events, in the early

BEGINNINGS OF QUANTUM MECHANICS / 51

part of 1926, that Schrödinger's wave-mechanics and the Göttingen quantum mechanics were mathematically equivalent. Since Schrödinger's method of partial differential equations was more familiar to physicists than the matrices, it could conveniently be used to work out matrix elements. Thus in Göttingen we used the summer of 1926 to get acquainted with Schrödinger's method, and this was most simply accomplished by writing a paper on a special physical problem, in which both methods could be studied in cooperation. Born wrote a paper on collision-processes, Jordan developed a general theory of transformation, and I myself attempted to calculate the spectrum of helium, and lighted in doing so on the relation between the symmetry of the wave-function under the permutation-group and the existence of the noncombining term-systems ortho- and parahelium. From the summer semester of 1926 I was working, in fact, in Copenhagen, but the links between the three centers of Göttingen, Copenhagen and Munich were at that time already so close that we quite regularly exchanged letters, and also met from time to time in one of the three cities.

All three papers already had a direct or indirect concern with the difficult problem of conceptual clarification. But before I discuss this in detail, I must mention two discussions that took place in the summer of 1926 between Schrödinger and the exponents of quantum mechanics. Sommerfeld had invited Schrödinger to Munich, so that in July he might lecture on his theory under the auspices of the colloquium there. Schrödinger at that time regarded his waves as true three-dimensional matter-waves—comparable, say, to electromagnetic waves—and wanted to eliminate entirely the discontinuous features of quantum theory, especially the so-called quantum jumps. To this I then protested in the discussion, since in this way, I thought, one would not even be able to explain Planck's law of thermal radiation. But no agreement could at that time be reached, and most of the other physicists hoped, as Schrödinger did, that the quantum jumps could somehow be avoided. In September detailed discussions then took place between Bohr and Schrödinger in Copenhagen, which lasted, as I remember, for more than a week, and in which I took part for as long as I could. This time there were passionate disputes, and the discussions were carried through to the bitter end. At the finish, we Copenhagen workers were convinced that Schrödinger's interpretation was untenable, and that quantum jumps were an essential part of the atomic process; and Schrödinger had certainly grasped that, at least initially, he was faced with insoluble difficulties.

Meanwhile Born, here in Göttingen, had achieved a crucial step forward. Following the Schrödinger proof of equivalence, in his theory of collision-processes, he was investigating the Schrödinger waves in many-dimensional configuration-space, not in the space of three dimensions. He came to the conclusion that the square of this wavefunction had to be regarded as a measure for the probability of the configuration in question. This clearly implied that matter-waves in three-dimensional space do not permit an adequate description of nature, and that the quantum theory contains a statistical element. In connection with the Copenhagen discussion, I then went on to inquire, in a small paper, whether, in resonance between two atoms, the exchange of energy is continuous or discontinuous. Given the principles of quantum mechanics already established, this could be derived from the oscillation phenomena, and the result again turned out to be unambiguously in favor of discontinuity, and hence of the quantum jumps. Finally, Jordan's theory of general unitary transformations, which was further supported by a paper of Dirac's with a similar content, showed that the squares of the elements of the transformation matrix would have to be interpreted as the probability of a transition from the one configuration to the other. But even with these confirmations, complete conceptual enlightenment was not yet achieved; for we still did not know how we were to describe, in quantum mechanics, so simply observable a phenomenon as the path of an electron in the cloud-chamber.

In the months from about October 1926 to February 1927, this problem was discussed in Copenhagen almost without a break. Already in his first talks with me on the Hainberg at Göttingen, in the summer of 1922, Bohr had repeatedly pointed out, indeed, that the ordinary language of the physicist was plainly unequal to describing events within the atom. It was now a matter of discovering which concepts of that language had to be retained, and which given up. Bohr and I were seeking the answer to the riddle here in somewhat different directions. I had meanwhile been so far educated by the Göttingen mathematical school as to assume that, through logical application of the quantum-mechanical formalism, conclusions must also be inferrable as to the remainder of the old concepts that would survive in the new language. But Bohr wanted to set out from the two initially contradictory pictures of the wave and the corpuscular theories, and to push on from thence to the correct concepts. The answer was then, as you know, made possible by reversing the statement of the problem; the question was no longer to be, "How do we represent the path of the electron in the cloud-chamber?"; instead, we had to ask, "Are there, perhaps, in the observation of nature, only such experimental situations as can be represented in the mathematical formalism of quantum theory?" Is it, in other words, correct, as Einstein once maintained against me, that theory first decides what can be observed? The answer could then be given in the form of the uncertainty-relation. The concept of path may be used only with the degree of inexactness characterized by the fact that the product of the uncertainty of position and the uncertainty of the associated momentum cannot be smaller than Planck's quantum of action. Bohr had arrived at the same limitations of language by way of the concept of complementarity formulated by him, and only now was it possible to state clearly what we are to understand by an observation-situation, and how it is represented in the mathematical formalism. Pauli at once concurred in this interpretation.

With this I can conclude my account of the beginnings of quantum mechanics in Göttingen. Important applications were already being appended to these beginnings in 1926 and 1927; I may mention as the most important the work of Hund and Wigner on the distribution of terms in symmetry classes. But from then on the development extends so rapidly that its depiction would far surpass the confines of this lecture. I would wish, therefore, to limit myself to the modest beginnings, and to spend only a little time in comparing contemporary physics with the physics of that era. It is so often said nowadays, that those days were a golden time for physics, when major discoveries could be made within a short period, whereas today the work has to go on laboriously and often in routine fashion. But I cannot accede to this for contemporary particle physics. There are certainly important differences, which are governed, for example, by the fact that all experiments on elementary particles are far more cumbersome than those of the old days on atomic shells, and that the mathematical forms, which could serve to describe the particle processes, have probably not yet been developed far enough by the mathematicians. But on the most important point, the two groups of problems are very similar. For in particle physics, too, there is the necessity of abandoning certain fundamental concepts of the earlier physics. Just as in relativity theory the old concept of simultaneity had to be sacrificed, and in quantum mechanics the notion of electron pathways, so in particle physics the concept of division, or of "consisting of," has to be given up. The history of physics in this century teaches us, that this abandonment of earlier concepts is much harder than the adoption of new ones. We shall have to reconcile ourselves to that. But I believe that the chance of achieving full clarity in the physics of elementary particles lies only with someone who, as in those days—if I may use, here, a more Eastern idiom—is in a position to make this sacrifice, not with the understanding only, but also with the heart. Such a development is just as interesting and exciting today as it was fifty years ago, and I hope that the new generation engages in it with just as much joy.

Cosmic Radiation and Fundamental Problems in Physics*

Cosmic-ray research has advanced our understanding of fundamental problems in physics, when concepts previously used are shown to have a limited range of applicability. Since cosmic rays contain information on the behavior of matter in the smallest (elementary particles) and largest dimensions (the universe), they have been particularly valuable in testing the concepts of daily life in relation to their meaning in physics, and in leading physicists to find new ones.

Cosmic radiation has—ever since its discovery some sixty years ago—played a very important role in the development of physics. It is a very interesting history which led from the first indication of rays coming from outer space to the earth to the discovery of very energetic particles in this ra-

---

*From 14th International Cosmic Ray Conference, Conference Papers, Vol. 11, pp. 3461–74, Max Planck Institute for Extra-terrestrial Physics, Munich, 1975; Die Naturwissenschaften, 63, pp. 63–67 (1976), Springer, 1976. (Lecture to the 14th International Cosmic Ray Conference on August 18, 1975 in Munich; written in English.)

diation, of new particles with unexpected properties, of new fundamental symmetries in the laws of nature, and finally to a great wealth of information about the residual matter and magnetic fields in interstellar space, and about those processes which can possibly produce the cosmic radiation. But I will not follow this historical line.

I will try to confine my talk to those fundamental problems of physics which have been touched or essentially advanced by the progress of knowledge in cosmic radiation. It is the interaction between this very special field of physics and the fundamental problems, basic to all physics, in which I shall be interested in this talk. This interaction became visible for the first time in the early thirties, when cosmic radiation played an essential part in one of the most important discoveries in physics of this century, the discovery of the positron. It is true that this discovery has not been made primarily in cosmic-ray research. Dirac in his theory of the electron had predicted a positively charged counterpart to the electron, but the first convincing evidence for its existence was found in the cosmic radiation by Anderson, and by Blackett and Occhialini. The first cloud-chamber pictures of showers, where photons created pairs of electron and positron, and these particles again created photons when traversing matter, demonstrated beyond any doubt the existence of the positrons and the validity of Dirac's theory. Shortly afterwards the positrons could also be seen in nuclear processes, that is in β-decay.

I should perhaps add a few words about the fundamental importance of this discovery. Up to that time, physicists had, perhaps more or less unconsciously, followed the philosophy of the ancient Greek philosopher Democritus. When one tries to divide a piece of matter over and over again, one would—so was the conjecture—eventually end up with smallest parts of matter which could not be divided any further, and therefore were called atoms. These atoms were taken as indivisible, unchangeable units of matter, as the building blocks from which all matter is constructed, and the atoms—or as we would nowadays say: elementary particles—should by their relative position and motion determine the visible properties of the various kinds of matter. This whole picture, plausible as it may seem, has been destroyed completely by the theory of Dirac and by its consequence, the discovery of the positron. The decisive point was not so much the existence of a new, hitherto unknown particle—many new particles have later been found without serious consequences for the foundations of physics—it was the discovery of a new symmetry, the particle-antiparticle conjugation, which was closely connected with the Lorentz group of special relativity, and with the transmutation from energy to matter and vice versa. In nonrelativistic physics the number of particles of any kind was a constant of motion like energy or momentum. In relativistic physics this number was not a good quantum number any more. A hydrogen atom, for example, did not necessarily consist of proton and electron, it may be taken as consisting of proton, two electrons and one positron, even if this latter configuration would only amount to a small relativistic correction of the complete wave function of hydrogen. One of the consequences of this situation was the conjecture that in a very energetic collision of two particles a larger number of new particles may be created, and these possibilities should be limited only by the laws of conservation of energy, momentum, isospin etc. It was again in cosmic radiation that this conjecture could be tested.

Actually already in the late thirties Blau and Wambacher had discovered in photographic plates, exposed at high altitudes, the so-called stars, events in which from the same point in the plate a great number of tracks started. Apparently an atomic nucleus had been hit by an incoming, very energetic particle and had emitted, as a result of the collision, a number of different particles. The interpretation of these stars was not simple, since the beginning of the process could possibly be a kind of cascade in the nucleus, similar to the well-known electron-positron cascades, followed by some evaporation of the nucleus. So these results did not immediately demonstrate the multiple production of particles in a collision of only two of them, which I mentioned as a conjecture. But in the course of time the cosmic-ray experiments could be refined, and after fifteen years the occurrence of multiple production was definitely established.

These results meant that the concepts of "dividing" and "consisting of" have only a limited range of applicability. Just as in relativity the concept "simultaneous" or in quantum theory the concepts "position" and "velocity" can be applied only with characteristic restrictions, and lose their meaning when used uncritically at the wrong place, so also the concepts "dividing" or "consisting of" are well defined only in special situations. When a particle, by a small amount of energy, can be disintegrated into two or several parts, the rest mass of which is very large compared with this small energy, then and only then may one say that the particle consists of these parts, can be divided into these parts. In all other cases the words "dividing" or "consisting of" have no well-defined meaning. What actually happens in a very energetic collision of two particles is the creation of new particles out of the kinetic energy. Energy becomes matter by assuming the form of elementary particles. But again the distinction between "elementary" particle and "compound system" has no well-defined meaning. Particles are stationary states of the physical system "matter." All these very important and fundamental results had their experimental basis in cosmic-ray research.

Another interesting result of cosmic-ray research was the discovery of the muon or μ-meson by Neddermayer and Anderson in 1937. This object was first mistaken as the particle which had been predicted by Yukawa as the material counterpart to the strong interaction between nucleons. But it soon turned out that the interaction of muons with heavy particles like proton and neutron was much too small; the muon could not be responsible for the strong interaction in a nucleus. Rather, the muon appeared as a heavier brother of the electron, different from it only by its larger mass. The discovery of the muon did not cause as fundamental a change in the basis of physics as the discovery of the positron. But it revealed an interesting feature in the spectrum of particles. This spectrum is divided into *two only weakly combining termsystems*, the hadrons and the leptons. Such weakly combining termsystems are well known from the optical spectra of atoms. But whether the causes for such splitting are similar in both cases is still an open question. The muons constitute—besides the neutrinos—the most penetrating part of the cosmic radiation and therefore play an important role in determining the intensity of the cosmic radiation as a function of height in the atmosphere.

I should perhaps mention another rather odd case in which the muons helped to settle a very fundamental question. In our country just before the war the theory of relativity was not accepted by the political power, and it was especially the dilatation of time in moving bodies, which was criticized as absurd and pure theoretical speculation. There were even trials concerning the question whether the theory of relativity could be taught at universities. In one of these discussions I could point out that the decay time of muons should depend on their velocity; muons which move almost with the velocity of light decay more slowly than those with smaller velocities—this was the prediction of the theory of relativity. The experimental results confirmed this prediction; the dilatation of time could be observed directly and the way was open for courses on relativity. So I have always felt grateful to the muons.

Shortly after the war, Powell in Bristol discovered the pion, which plays a very important role in most cosmic-ray phenomena. This object fulfills all the conditions formulated

COSMIC RADIATION AND PHYSICS / 61

by Yukawa for the material counterpart of strong interaction; it was, as was recognized later, not the only particle of this kind, but being the hadron of the smallest mass it was soon discovered in almost all events of very high energy. Besides that, the pion decays into muon and neutrino, so the origin of the muons had been clarified.

Like the muons, the pion did not cause fundamental changes in the basis of physics. It just confirmed that the various particles are stationary states of the system matter, different through their different behavior under the transformation of the fundamental group. The groups are more fundamental than the particles.

At that time, besides the Lorentz group of relativity, only the isospin group was known as fundamental. It had been found in 1932 in connection with nuclear physics; but it was first through the pions that its fundamental character was fully understood. Cosmic-ray experiments on the pion demonstrated that the isospin group is an exact symmetry for the strong interaction, and only the electromagnetic interaction and weaker interactions break this symmetry. These results could be interpreted by assuming that the natural law underlying the spectrum of particles is exactly invariant under the isospin transformation, and that the deviations from this symmetry are caused by an asymmetric, degenerate ground state. Similar situations are well known from quantum mechanics of solid bodies.

Almost simultaneously with the pion other particles were discovered in the cosmic radiation—heavier than the pion and somewhat "strange" in their behavior. They had a rather long lifetime of the order 10^{-10} sec and therefore their tracks could be observed in cloud chambers or in emulsions. But this long lifetime could not be understood if only the known symmetries and corresponding quantum numbers (baryonic number, isospin, angular momentum) were considered—one would have expected a very much shorter lifetime—and insofar their behavior was strange.

The correct interpretation was given by Pais in 1952, when he introduced a new quantum number, called strangeness, and the corresponding symmetry (or transformation property). So cosmic-ray research had led to a new symmetry group; and since, as I mentioned before, groups are more important than particles, this was again a very essential contribution to fundamental problems in physics.

There was general agreement among most physicists at that time that many more particles could be found, if objects with very short lifetime could be made observable. The particles are just stationary states of the system matter, therefore one had to expect many different particles, most of them of very short lifetime. Such objects could be observed only as so-called *resonance states,* and for that purpose better statistics were needed than cosmic-ray observations could offer. Fortunately for the particle physicists, the first big accelerators had been built at that time and went into action, the cosmotron in Brookhaven, the bevatron in Berkeley and the proton synchrotron of CERN in Geneva. So for a long time to come, the important results in particle physics were obtained with the help of big accelerators, while cosmic-ray research turned its attention primarily to astrophysical problems. This development was unavoidable, but did not always meet the wishes of the particle physicists; it was a sad change.

The romantic time had gone, in which the study of cloud-chamber pictures in a mountain laboratory at high altitude could be combined with skiing and mountaineering, or in which balloon experiments could be started from a beautiful island in the Mediterranean with the help of an airplane and a military vessel from the Italian navy, as had been managed by our Italian friends. Certainly the warm sun of the Mediterranean has contributed to the scientific success of the experiments. But this gay time had now gone, and particle research had to be done in the "matter of fact" atmosphere of huge accelerator establishments.

COSMIC RADIATION AND PHYSICS / 63

In astrophysics, cosmic radiation became a valuable new tool which promised information beyond that obtained from the visible or infrared light of the stars. The first problem was of course the origin of the radiation. Already Forbush recognized that some low-energy part of the cosmic radiation was occasionally emitted by the sun, by some turbulent phenomena at its surface. But it was understood very soon that a definite answer to the problem of the origin of cosmic rays required a thorough knowledge of the electromagnetic fields in the plasma between the stars, in our planetary system—here I may remind you of the solar wind, first discussed by Biermann—in our galaxy and finally in extragalactic space. Investigations of these fields has become a central part of astrophysics in recent years, and much information has been obtained with the help of the cosmic radiation. Concerning its origin the general opinion seems to be now that supernovae and their relics, the pulsars, are the main sources of high-energy cosmic rays. But I should not go into the details of astrophysics, but rather come back to my first question: where does cosmic radiation touch fundamental problems of physics?

I just mentioned the pulsars, which belong to the stars of highest density hitherto observed. Their matter density is comparable to the density of an atomic nucleus. They are held together by gravitational forces. Such stars raise two fundamental problems; one concerns the relation of gravitation to the other forces of interaction within matter, the other concerns the equation of state for matter of that—or still higher—density. But before I come to these problems, I would like to mention a few valuable contributions of cosmic-ray research to very important problems in particle physics, even during the time of the big accelerators.

The particles of the cosmic radiation have energies up to 10^{19}e V, and it is obvious that such high energies cannot be reached by accelerators, at least not in the near future. Hence the collisions of particles at such extreme energies could be investigated only in the cosmic radiation—and even if low intensity and poor statistics prevented accurate results, the question was put how the cross-section or other characteristics of showers should change with energy in the range of extremely high energies. Is there an asymptotic region, far above the energies of the common particles or resonance states, where no more spectacular new events or drastic changes are found or expected? The information obtained from the cosmic radiation on this problem was scarcely more than a vague indication; still, it stimulated theoretical investigations, which, more than twenty years ago, led to the conjecture, that the total cross-section for collisions between any hadrons should at high energies increase with the square of the logarithm of the energy. Hence there should be an asymptotic region; but the total cross-sections in this region should not be constant, they should increase logarithmically. This conjecture has been borne out in recent experiments with the storage rings in CERN and with the Batavia accelerator. The asymptotic region seems to start at an energy of the order of 10 GeV in the center of mass system, and has been followed up to 50 GeV for proton-proton collisions in the storage rings in CERN. The essential contribution from the Batavia machine was the result that the logarithmic increase could be observed also for collisions of pions or kaons with protons. This was a strong argument in favor of the assumption that there is a general asymptotic region and that it had been reached in these experiments. For the purpose of understanding this asymptotic region it is sufficient to describe the particles as nearly spherical clouds of continuous matter without any reference to particles, of which these clouds could consist. This is satisfactory since the word "consisting of" has as a rule lost its meaning in particle physics.

Another problem has occupied the minds of the particle physicists during the last ten years. We know that the group SU_3 plays a role in the spectrum of particles as an approximate symmetry. The simplest representation of SU_3 is three-dimensional, therefore one could expect a triplet of particles corresponding to this representation; the electric charge of these particles would be ⅓ or ⅔ of the elementary charge, and they have been given the name "quarks." However, such particles have never been observed in experiments with the big machines. Therefore it was suggested that the quarks might be rather heavy, held together by very big binding energies, so that the existing machines would not be sufficient to take them apart. At this point cosmic radiation turned out to be very helpful, because the energy of primaries in the cosmic radiation may be a thousand or more times bigger than the maximum particle energy in a big machine. The fact that even in the cosmic radiation no quarks have been found, is a very strong argument in favor of their non-existence. If such results are final, it seems to me very difficult to give any well defined meaning to the statement, "a proton consists of three quarks," because neither the word "consists of" nor the word "quark" have a well defined meaning. How could such a sentence then be interpreted? The same scepticism is justified with respect to other particles, which have been predicted, but not found: W-mesons, partons, gluons, magnetic poles, charmed particles. If they cannot be observed, either in big machines or in the cosmic radiation, it is difficult to argue that they are good concepts in a phenomenological description. Here we meet a situation which is well known already from quantum mechanics. Our normal language induces us to ask questions which have no meaning; for example, "What is the orbit of an electron moving around an atomic nucleus?" Neither the word "orbit" nor the word "moving" are well defined on account of the relations of uncertainty; hence the question has no meaning.

This leads me to a central problem, which is closely connected to experiences in cosmic radiation. But before I discuss the empirical side, I would like to explain its fundamental importance in particle physics and physics in general.

From the experiments of the last decades we have learned that the different particles are just different stationary states of the system matter. They are characterized by quantum numbers, or, if you prefer, by their transformation properties under fundamental groups. Theoretical understanding of particle physics can only mean an understanding of the spectrum of particles. A single line in the optical spectrum of iron cannot be understood; but the spectrum can be understood, it can be reduced to the Schrödinger equation of a system, containing twenty-six electrons and the iron nucleus.

The essential elements of the theoretical interpretation of a spectrum are well known and can be learned both from classical physics and from quantum mechanics. We can think of the elastic vibrations of a string, or of the electromagnetic vibrations in a cavity, or of the stationary states of an atom, for example, the iron atom. In all cases we first need a precise statement about the dynamical properties of the system, and then we have to add the special boundary conditions. In the case of the string a precise mathematical formulation of the elastic and dynamic properties of the string is the first step; then, by stating where the string is fixed, we can calculate the spectrum of the vibrations. For the electromagnetic vibrations in a cavity, Maxwell's equations define the dynamical properties of the system. The boundary conditions are given by the form of the cavity. On account of the complexity of the problem it will frequently not be possible to calculate the whole spectrum precisely; but for the lowest vibrations one should be able to get good approximations. For the iron atom the dynamical properties are defined by quantum mechanics, that is, the Schrödinger equation. By the supplementary condition that the wave-function must vanish at infinity, the stationary states are fixed. If the atom were to be enclosed in a small box, the stationary states would be different.

Starting from these analogies it is clear that the first condition for understanding the particle spectrum is a precise mathematical formulation of the dynamics of matter. It is obvious that the word particle should not enter in this formulation. Because the particles are defined later on by combining the dynamics of the system matter with the boundary conditions, the particles are secondary structures. The spectrum of the particles may easily be different in our surroundings in the universe from that in the interior of a very dense neutron star, because the boundary conditions in both cases may be different. From this you see the fundamental importance of the dynamics of matter, and the question arises, how we can get hold of its mathematical formulation.

Since the particle concept is not useful in this connection, the group properties of the dynamical law must play a decisive role. The dynamical law of the vibrating string, for example, is invariant under translations in time and translations along the string, and under rotations around the string. By the boundary conditions the second invariance is broken, the third as a rule is not broken. For the electromagnetic vibrations in a cavity the dynamical law is invariant under the full Lorentz group: the invariance is partially broken by the boundary conditions.

For the dynamics of matter some of the essential invariances are known: the Lorentz group and the isospin group SU_2. Also the scale group should possibly be counted under the fundamental invariances. But I should not go into the details of these symmetries of the dynamical law. I would rather like to come back to cosmic radiation. How can research in cosmic radiation or more generally in astrophysics contribute to our knowledge of the dynamics of matter?

First a word about causality. We know from the dispersion relations that interaction in matter follows the law of causality. The exact mathematical formulation of this statement is perhaps not completely known, but we have good reason to believe that interaction can be formulated as local interaction, as it is, for example, in quantum electrodynamics. The non-local Coulomb force is compatible with this statement. Starting from such a situation it is plausible that the study of matter of extremely high density should give the most direct information about this local interaction, and thereby about the dynamics of matter.

In a neutron star the density is of the same order as in an atomic nucleus. At such densities it still has meaning to say that the nucleus consists of a number of nucleons, because a small amount of energy—small compared with the rest mass of a nucleus—is sufficient to take a proton or a neutron out of the nucleus. The nucleons are still far enough apart from each other in the nucleus—their energy of interaction therefore is small compared to their rest mass. This is equally true in a neutron star, and therefore it has been possible to get estimates for the equation of state of such stellar matter. If, however, the density is considerably higher—e.g., in a star of larger mass contracted by gravitation—then the question of which particles the star consists has no well-defined meaning. The space available for a particle would be smaller than its normal size, therefore it could not have its normal mass; the interaction is so strong, that particles would as a rule not be on their mass-shell. In other words, one could only speak of a mixture of all particles, and then it is more reasonable to speak of continuous matter. It is the dynamical behavior of this continuous matter which is the fundamental problem in particle physics.

If it should become possible to get more information about the equation of state, not only in neutron stars, but especially in stars of still higher density, this would be extremely important for understanding the dynamical behavior of matter. Whether observations in cosmic radiation or

COSMIC RADIATION AND PHYSICS / 69

in wider fields of astrophysics will be more helpful I cannot judge. I only wanted to emphasize the importance of the problem.

There is one other special field in cosmic radiation where this problem of the dynamics of matter can be tackled from an entirely different side. If two particles of extremely high energy collide, then, in the first moment of the collision, one has a small disc of extremely dense matter, which then explodes and, by diminishing its density, finally disintegrates into many particles. This is the well-known process of multiple production of particles which is, of course, the more interesting the higher the energy of the colliding particles has been. If the primary cosmic ray particle has an energy of 10^6 GeV, then the density of the initial disc in the collision can be a thousand times greater than in a neutron star.

The study of the behavior of such cosmic-ray showers of extreme energy should therefore give valuable information on the dynamics of matter. It is encouraging in this connection that already in the storage rings of CERN and in the Batavia machine one seems to have reached, or at least approached, the asymptotic region. For the initial phase of collisions in this region the primary particles can simply be pictured as clouds of continuous matter the density of which falls off exponentially at the surface. This model explains the logarithmic increase of the total cross-section as a function of increasing energy. I should still point to a characteristic difference between the two experiments, those on the stars of extreme density and those on the discs after the collision of very energetic particles. In the first case gravitation plays an important role, in the second it is unimportant. Therefore the two kinds of experiments can give two different kinds of relevant information.

Coming back in conclusion to the general questions mentioned in the beginning of my talk I should perhaps say that the special role of cosmic radiation in the whole field of physics rests on two facts. Cosmic radiation contains information on the behavior of matter in the smallest dimensions and also it contributes to our knowledge about the structure of the universe—of the world—in the largest dimensions. These two extreme ends are not accessible to direct observation—they can be investigated only by very indirect deductions, in which the concepts of daily life have to be replaced by other rather abstract new concepts; only then will we learn, what such words as "extreme ends" or "infinity" can mean in relation to nature. In this sense cosmic radiation—in spite of any changes in the style of the experiments—can still be called a very romantic, a very inspiring science.

What Is an Elementary Particle?*

The question "What is an elementary particle?" must naturally be answered above all by experiment. So I shall first summarize briefly the most important experimental findings of elementary particle physics during the last fifty years, and will try to show that, if the experiments are viewed without prejudice, the question alluded to has already been largely answered by these findings, and that there is no longer much for the theoretician to add. In the second part I will then go on to enlarge upon the philosophical problems connected with the concept of an elementary particle. For I believe that certain mistaken developments in the theory of elementary particles—and I fear that there are such—are due to the fact that their authors would claim that they do not wish to trouble about philosophy, but that in reality they unconsciously start out from a bad philosophy, and have therefore fallen through preju-

---

*From a lecture to the Session of the German Physical Society on March 5, 1975. In *Die Naturwissenschaften*, 63, pp. 1–7 (1976), Springer, 1976.

dice into unreasonable statements of the problem. One may say, with some exaggeration, perhaps, that good physics has been inadvertently spoiled by bad philosophy. Finally I shall say something of these problematic developments themselves, compare them with erroneous developments in the history of quantum mechanics in which I was myself involved, and consider how such wrong turnings can be avoided. The close of the lecture should therefore be more optimistic again.

First, then, to the experimental facts. Not quite fifty years ago, Dirac, in his theory of electrons, predicted that in addition to electrons there would also have to exist the appropriate anti-particles, the positrons; and a few years later the existence of positrons, their origin in pair-creation, and hence the existence of so-called anti-matter, was experimentally demonstrated by Anderson and Blackett. It was a discovery of the first order. For till then it had mostly been supposed that there are two kinds of fundamental particle, electrons and protons, which are distinguished above all others by the fact that they can never be changed, so that their number is always constant as well, and which for that very reason had been called elementary particles. All matter was supposed in the end to be made up of electrons and protons. The experimental proof of pair-creation and positrons showed that this idea was false. Electrons can be created and again disappear; so their number is by no means constant; they are not elementary in the sense previously assumed.

The next important step was Fermi's discovery of artificial radioactivity. It was learnt from many experiments that one atomic nucleus can turn into another by emission of particles, if the conservation laws for energy, angular momentum, electric charge, etc., allow this. The transformation of energy into matter, which had already been recognized as possible in Einstein's relativity theory, is thus a very commonly observable phenomenon. There is no talk here of any conservation of the number of particles. But there are indeed physical properties, characterizable by quantum numbers—I am thinking, say, of angular momentum or electric charge—in which the quantum numbers can then take on positive and negative values, and for these a conservation law holds.

In the thirties there was yet another important experimental discovery. It was found that in cosmic radiation there are very energetic particles, which, on collision with other particles, say a proton, in the emulsion of a photographic plate, can let loose a shower of many secondary particles. Many physicists believed for a time that such showers can originate only through a sort of cascade formation in atomic nuclei; but it later turned out that, even in a collision between a single pair of energetic particles, the theoretically conjectured multiple production of secondary particles does in fact occur. At the end of the forties, Powell discovered the pions, which play the major part in these showers. This showed that in collisions of high-energy particles, the transformation of energy into matter is quite generally the decisive process, so that it obviously no longer makes sense to speak of a splitting of the original particle. The concept of "division" had come, by experiment, to lose its meaning.

In the experiments of the fifties and sixties, this new situation was repeatedly confirmed: many new particles were discovered, with long and short lives, and no unambiguous answer could be given any longer to the question about what these particles consisted of, since this question no longer has a rational meaning. A proton, for example, could be made up of neutron and pion, or Λ-hyperon and kaon, or out of two nucleons and an anti-nucleon; it would be simplest of all to say that a proton just consists of continuous matter, and all these statements are equally correct or equally false. The difference between elementary and composite particles has thus basically disappeared. And that is no doubt the most important experimental finding of the last fifty years.

As a consequence of this development, the experiments have strongly suggested an analogy: the elementary particles are something like the stationary states of an atom or a molecule. There is a whole spectrum of particles, just as there is a spectrum, say, of the iron atom or a molecule, where we may think, in the latter case, of the various stationary states of a molecule, or even of the many different possible molecules of chemistry. Among particles, we shall speak of a spectrum of "matter." In fact, during the sixties and seventies, the experiments with the big accelerators have shown that this analogy fits all the findings so far. Like the stationary states of the atom, the particles, too, can be characterized by quantum numbers, that is, by symmetry- or transformation-properties, and the exact or approximately valid conservation principles associated with them decide as to the possibility of the transformations. Just as the transformation properties of an excited hydrogen atom under spatial rotation decide whether it can fall to a lower state by emission of a photon, so too, the question whether a ϕ-boson, say, can degenerate into a ρ-boson by emission of a pion, is decided by such symmetry properties. Just as the various stationary states of an atom have very different lifetimes, so too with particles. The ground state of an atom is stable, and has an infinitely long lifetime, and the same is true of such particles as the electron, proton, deuteron, etc. But these stable particles are in no way more elementary than the unstable ones. The ground state of the hydrogen atom follows from the same Schrödinger equation as the excited states do. Nor are the electron and photon in any way more elementary than, say, a Λ-hyperon.

The experimental particle-physics of recent years has thus fulfilled much the same tasks, in the course of its development, as the spectroscopy of the early twenties. Just as, at that time, a large compilation was brought out, the so-called

WHAT IS AN ELEMENTARY PARTICLE? / 75

Paschen-Götze tables, in which the stationary states of all atom shells were collected, so now we have the annually supplemented *Reviews of Particle Properties*, in which the stationary states of matter and its transformation-properties are recorded. The work of compiling such a comprehensive tabulation therefore corresponds, say, to the star-cataloging of the astronomers, and every observer hopes, of course, that he will one day find a particularly interesting object in his chosen area.

Yet there are also characteristic differences between particle physics and the physics of atomic shells. In the latter we are dealing with such low energies, that the characteristic features of relativity theory can be neglected, and nonrelativistic quantum mechanics used, therefore, for description. This means that the governing symmetry-groups may differ in atomic-shell physics on the one hand, and in particles on the other. The Galileo group of shell physics is replaced, at the particle level, by the Lorentz-group; and in particle physics we also have new groups, such as the isospin group, which is isomorphic to the SU_2 group, and then the SU_3 group, the group of scaling transformations and still others. It is an important experimental task to define the governing groups of particle physics, and in the past twenty years it has already been largely accomplished.

Here we can learn from shell physics, that in those very groups which manifestly designate only approximately valid symmetries, two basically different types may be distinguished. Consider, say, among optical spectra, the O_3 group of spatial rotations, and the $O_3 \times O_3$ group, which governs the multiplet-structure in spectra. The basic equations of quantum mechanics are strictly invariant with respect to the group of spatial rotations. The states of atoms having greater angular momenta are therefore severely degenerate, that is, there are numerous states of exactly equal energy. Only if the atom is placed in an external electromagnetic field do the states split up, and the familiar fine structure emerge, as in the Zeeman or Stark effect. This degeneracy can also be abolished if the ground state of the system is not rotation-invariant, as in the ground states of a crystal or ferro-magnet. In this case there is also a splitting of levels; the two spin-directions of an electron in a ferro-magnet are no longer associated with exactly the same energy. Furthermore, by a well-known theorem of Goldstone, there are bosons whose energy tends to zero with increasing wavelength, and, in the case of the ferro-magnet, Bloch's spin-waves or magnons.

It is different with the group $O_3 \times O_3$, from which result the familiar multiplets of optical spectra. Here we are dealing with an approximate symmetry, which comes about in that the spin-path interactions in a specific region are small, so that the spins and paths of electrons can be skewed counter to each other, without producing much change in the interaction. The $O_3 \times O_3$ symmetry is therefore also a useful approximation only in particular parts of the spectrum. Empirically, the two kinds of broken symmetry are most clearly distinguishable in that, for the fundamental symmetry broken by the ground state, there must, by the Goldstone theorem, be associated bosons of zero rest mass, or long-range forces. If we find them, there is reason to believe that the degeneracy of the ground state plays an important role here.

Now if these findings are transferred from atomic-shell physics to particle physics, it is very natural, on the basis of the experiments, to interpret the Lorentz group and the SU_2 group, the isospin group, that is, fundamental symmetries of the underlying law of nature. Electromagnetism and gravitation then appear as the long-range forces associated with symmetry broken by the ground state. The higher groups, SU_3, SU_4, SU_6, or $SU_2 \times SU_2$, $SU_3 \times SU_3$ and so on, would then have to rank as dynamic symmetries, just like $O_3 \times O_3$ in atomic-shell physics. Of the dilatation or scaling group, it may be doubted whether it should be counted among the fundamental symmetries; it is perturbed by the existence of particles with finite mass, and by the gravitation due to masses in the universe. Owing to its close relation to the Lorentz group, it certainly ought to be numbered among the fundamental symmetries. The foregoing assignment of perturbed symmetries to the two basic types is made plausible, as I was already saying, by the experimental findings, but it is not yet possible, perhaps, to speak of a final settlement. The most important thing is that, with regard to the symmetry groups that present themselves in the phenomenology of spectra, the question must be asked, and if possible answered, as to which of the two basic types they belong to.

Let me point to yet another feature of shell-physics: among optical spectra there are non-combining, or more accurately, weakly combining term-systems, such as the spectra of para- and ortho-helium. In particle physics we can perhaps compare the division of the fermion spectrum into baryons and leptons with features of this type.

The analogy between the stationary states of an atom or molecule, and the particles of elementary particle physics, is therefore almost complete, and with this, so it seems to me, I have also given a complete qualitative answer to the initial question "What is an elementary particle?" But only a qualitative answer! The theorist is now confronted with the further question, whether he can also underpin this qualitative understanding by means of quantitative calculations. For this it is first necessary to answer a prior question: What is it, anyway, to understand a spectrum in quantitative terms?

For this we have a string of examples, from both classical physics and quantum mechanics alike. Let us consider, say, the spectrum of the elastic vibrations of a steel plate. If we are not to be content with a qualitative understanding, we shall start from the fact that the plate can be characterized by specific elastic properties, which can be mathematically represented. Having achieved this, we still have to append the boundary conditions, adding, for example, that the plate is circular or rectangular, that it is, or is not, under tension, and from this, at least in principle, the spectrum of elastic or acoustic vibrations can be calculated. Owing to the level of complexity, we shall certainly not, indeed, be able to work out all the vibrations exactly, but may yet, perhaps, calculate the lowest, with the smallest number of nodal lines.

Thus two elements are necessary for quantitative understanding: the exactly formulated knowledge, in mathematical terms, of the dynamic behavior of the plate, and the boundary conditions, which can be regarded as "contingent," as determined, that is, by local circumstances; the plate, of course, could also be dissected in other ways. It is like this, too, with the electrodynamic oscillations of a cavity resonator. The Maxwellian equations determine the dynamic behavior, and the shape of the cavity defines the boundary conditions. And so it is, also, with the optical spectrum of the iron atom. The Schrödinger equation for a system with a nucleus and 26 electrons determines the dynamic behavior, and to this we add the boundary conditions, which state in this instance that the wave-function shall vanish at infinity. If the atom were to be enclosed in a small box, a somewhat altered spectrum would result.

If we transfer these ideas to particle physics, it becomes a question, therefore, of first ascertaining by experiment the dynamical properties of the matter system, and formulating this in mathematical terms. As the contingent element, we now add the boundary conditions, which here will consist essentially of statements about so-called empty space, i.e., about the cosmos and its symmetry properties. The first step must in any case be the attempt to formulate mathematically a law of nature that lays down the dynamics of matter. For the second step, we have to make statements about the boundary conditions. For without these, the spectrum just cannot be defined. I would guess, for example, that in one of the "black holes" of contemporary astrophysics, the spectrum of elementary particles would look totally different from our own. Unfortunately, we cannot experiment on the point.

But now a word more about the decisive first step, namely the formulation of the dynamical law. There are pessimists among particle physicists, who believe that there simply is no such law of nature, defining the dynamic properties of matter. With such a view I confess that I can make no headway at all. For somehow the dynamics of matter has to exist, or else there would be no spectrum; and in that case we should also be able to describe it mathematically. The pessimistic view would mean that the whole of particle physics is directed, eventually, at producing a gigantic tabulation containing the maximum number of stationary states of matter, transition-probabilities and the like, a "Super-Review of Particle Properties," and thus a compilation in which there is nothing more to understand, and which therefore, no doubt, would no longer be read by anyone. But there is also not the least occasion for such pessimism, and I set particular store by this assertion. For we actually observe a particle spectrum with sharp lines, and so, indirectly, a sharply defined dynamics of matter as well. The experimental findings, briefly sketched above, also contain already very definite indications as to the fundamental invariance properties of this fundamental law of nature, and we know from the dispersion relations a great deal about the level of causality that is formulated in this law. We thus have the essential determinants of the law already to hand, and after so many other spectra in physics have finally been understood to some extent in quantitative terms, it will also be possible here, despite the high degree of complexity involved. At this point—and just because of its complexity—I would sooner not discuss the special proposal that was long ago made by myself, together with Pauli, for a mathematical formulation of the underlying law, and which, even now, I still believe to have the best chances of being the right one. But I would like to point out with all emphasis, that the formulation of such a law is the indispensable precondition for understanding the spectrum of elementary particles. All else is not understanding; it is hardly more than a start to the tabulation project, and as theorists, at least, we should not be content with that.

I now come to the philosophy by which the physics of elementary particles is consciously or unconsciously guided. For two and a half millennia, the question has been debated by philosophers and scientists, as to what happens when we try to keep on dividing up matter. What are the smallest constituent parts of matter? Different philosophers have given very different answers to this question, which have all exerted their influence on the history of natural science. The best known is that of the philosopher Democritus. In attempting to go on dividing, we finally light upon indivisible, immutable objects, the atoms, and all materials are composed of atoms. The position and motions of the atoms determine the quality of the materials. In Aristotle and his medieval successors, the concept of minimal particles is not so sharply defined. There are, indeed, minimal particles here for every kind of material—on further division the parts would no longer display the characteristic properties of the material—but these minimal parts are continuously changeable, like the materials themselves. Mathematically speaking, therefore, materials are infinitely divisible; matter is pictured as continuous.

The clearest opposing position to that of Democritus was adopted by Plato. In attempting continual division we ultimately arrive, in Plato's opinion, at mathematical forms: the regular solids of stereometry, which are definable by their symmetry properties, and the triangles from which they can be constructed. These forms are not themselves matter, but they shape it. The element earth, for example, is based on the shape of the cube, the element fire on the shape of the tetrahedron. It is common to all these philosophers, that they wish in some way to dispose of the antinomy of the infinitely small, which, as everyone knows, was discussed in detail by Kant.

Of course, there are and have been more naïve attempts at rationalizing this antinomy. Biologists, for example, have developed the notion that the seed of an apple contains an invisibly small apple tree, which in turn bears blossom and fruit; that again in the fruit there are seeds, in which once more a still tinier apple tree is hidden, and so *ad infinitum*. In the same way, in the early days of the Bohr-Rutherford theory of the atom as a miniature planetary system, we developed with some glee the thesis that upon the planets of this system, the electrons, there are again very tiny creatures living, who build houses, cultivate fields and do atomic physics, arriving once more at the thesis of their atoms as miniature planetary systems, and so *ad infinitum*. In the background here, as I said already, there is always lurking the Kantian antinomy, that it is very hard, on the one hand, to think of matter as infinitely divisible, but also difficult, on the other, to imagine this division one day coming to an enforced stop. The antinomy, as we know, is ultimately brought about by our erroneous belief that we can also apply our intuition to situations on the very small scale. The strongest influence on the physics and chemistry of recent centuries has undoubtedly been exerted by the atomism of Democritus. It permits an intuitive description of small-scale chemical processes. The atoms can be compared to the mass-points of Newtonian mechanics, and such a comparison leads to a satisfying statistical theory of heat. The chemist's atoms were not, indeed, mass-points at all, but miniature planetary systems, and the atomic nucleus was composed of protons and neutrons, but electrons, protons, and eventually even neutrons could, it was thought, quite well be regarded as the true atoms, that is, as the ultimate indivisible building-blocks of matter. During the last hundred years, the Democritean idea of the atom had thus become an integrating component of the physicist's view of the material world; it was readily intelligible and to some extent intuitive, and determined physical thinking even among physicists who wanted to have nothing to do with philosophy. At this point I should now like to justify my suggestion, that today in the physics of elementary particles, good physics is unconsciously being spoiled by bad philosophy.

We cannot, of course, avoid employing a language that stems from this traditional philosophy. We ask, "What does the proton consist of? Can one divide the electron, or is it indivisible?" "Is the light-quantum simple, or is it composite?" But these questions are wrongly put, since the words *divide* or *consist of* have largely lost their meaning. It would thus be our task to adapt our language and thought, and hence also our scientific philosophy, to this new situation engendered by the experiments. But that, unfortunately, is very difficult. The result is that false questions and false ideas repeatedly creep into particle physics, and lead to the erroneous developments of which I am about to speak. But first a further remark about the demand for intuitability.

There have been philosophers who have held intuitability to be the precondition for all true understanding. Thus here in Munich, for example, the philosopher Hugo Dingler has championed the view that intuitive Euclidean geometry is the only true geometry, since it is presupposed in the construction of our measuring instruments; and on the latter point, Dingler is quite correct. Hence, he says, the experimental findings which underlie the general theory of relativity should be described in other terms than those of a more general Riemannian geometry, which deviates from the Euclidean; for otherwise we become involved in contradictions. But this demand is obviously extreme. To justify what we do by way of experiment, it is enough that, in the dimensions of our apparatus, the geometry of Euclid holds

WHAT IS AN ELEMENTARY PARTICLE? / 83

to a sufficiently good approximation. We must therefore come to agree that the experimental findings on the very small and very large scale no longer provide us with an intuitive picture, and must learn to manage there without intuitions. We then recognize, for example, that the aforementioned antinomy of the infinitely small is resolved, among elementary particles, in a very subtle fashion, in a way that neither Kant nor the ancient philosophers could have thought of, namely inasmuch as the term *divide* loses its meaning.

If we wish to compare the findings of contemporary particle physics with any earlier philosophy, it can only be with the philosophy of Plato; for the particles of present-day physics are representations of symmetry groups, so the quantum theory tells us, and to that extent they resemble the symmetrical bodies of the Platonic view.

But our purpose here was to occupy ourselves not with philosophy, but with physics, and so I will now go on to discuss that development in theoretical particle physics, which in my view sets out from a false statement of the problem. There is first of all the thesis, that the observed particles, such as protons, pions, hyperons and many others, are made up of smaller unobserved particles, the quarks, or else from partons, gluons, charmed particles, or whatever these imagined particles may all be called. Here the question has obviously been asked, "What do protons consist of?" But it has been forgotten in the process, that the term *consist of* only has a halfway clear meaning if we are able to dissect the particle in question, with a small expenditure of energy, into constituents whose rest mass is very much greater than this energy-cost; otherwise, the term *consist of* has lost its meaning. And that is the situation with protons. In order to demonstrate this loss of meaning in a seemingly well-defined term, I cannot forebear from telling a story that Niels Bohr was wont to retail on such occasions. A small boy comes into a shop with twopence in his hand, and tells the shopkeeper that he would like twopence-worth of mixed sweets. The shopkeeper hands him two sweets, and says: "You can mix them for yourself." In the case of the proton, the concept "consist of" has just as much meaning as the concept of "mixing" in the tale of the small boy.

Now many will object to this, that the quark hypothesis has been drawn from empirical findings, namely the establishing of the empirical relevance of the SU_3 group; and furthermore, it holds up in the interpretation of many experiments on the application of the SU_3 group as well. This is not to be contested. But I should like to put forward a counter-example from the history of quantum mechanics, in which I myself was involved; a counter-example which clearly displays the weakness of arguments of this type. Prior to the appearance of Bohr's theory, many physicists maintained that an atom must be made up of harmonic oscillators. For the optical spectrum certainly contains sharp lines, and they can only be emitted by harmonic oscillators. The charges on these oscillators would have to correspond to other electromagnetic values than those on the electron, and there would also have to be very many oscillators, since there are very many lines in the spectrum.

Regardless of these difficulties, Woldemar Voigt constructed at Göttingen in 1912 a theory of the anomalous Zeeman effect of the D-lines in the optical spectrum of sodium, and did so in the following way: he assumed a pair of coupled oscillators which, in the absence of an external magnetic field, yielded the frequencies of the two D-lines. He was able to arrange the coupling of the oscillators with one another, and with the external field, in such a way that, in weak magnetic fields, the anomalous Zeeman effect came out correct, and that in very strong magnetic fields the Paschen-Back effect was also correctly represented. For the intermediate region of moderate fields, he obtained, for the frequencies and intensities, long and complex quadratic roots; formulae, that is, which were largely incomprehensible, but which obviously reproduced the experiments with great exactness. Fifteen years later, Jordan and I took the trouble to work out the same problem by the methods of the quantum-mechanical theory of perturbation. To our great astonishment, we came out with exactly the old Voigtian formulae, so far as both frequencies and intensities were concerned and this, too, in the complex area of the moderate fields. The reason for this we were later well able to perceive; it was a purely formal and mathematical one. The quantum-mechanical theory of perturbation leads to a system of coupled linear equations, and the frequencies are determined by the eigen values of the equation-system. A system of coupled harmonic oscillators leads equally, in the classical theory, to such a coupled linear equation-system. Since, in Voigt's theory, the most important parameter had been cancelled out, it was therefore no wonder that the right answer emerged. But the Voigtian theory contributed nothing to the understanding of atomic structure.

Why was this attempt of Voigt's so successful on the one hand, and so futile on the other? Because he was only concerned to examine the D-lines, without taking the whole line-spectrum into account. Voigt had made phenomenological use of a certain aspect of the oscillator hypothesis, and had either ignored all the other discrepancies of this model, or deliberately left them in obscurity. Thus he had simply not taken his hypothesis in real earnest. In the same way, I fear that the quark hypothesis is just not taken seriously by its exponents. The questions about the statistics of quarks, about the forces that hold them together, about the particles corresponding to these forces, about the reasons why quarks never appear as free particles, about the pair-creation of quarks in the interior of the elementary particle—all these questions are more or less left in obscurity. If there was a desire to take the quark hypothesis in real earnest, it would be necessary to make a precise mathematical approach to the dynamics of quarks, and the forces that hold them together, and to show that, qualitatively at least, this approach can reproduce correctly the many different features of particle physics that are known today. There should be no question in particle physics to which this approach could not be applied. Such attempts are not known to me, and I am afraid, also, that every such attempt which is presented in precise mathematical language would be very quickly refutable. I shall therefore formulate my objections in the shape of questions: "Does the quark hypothesis really contribute more to understanding of the particle spectrum, than the Voigtian hypothesis of oscillators contributed, in its day, to understanding of the structure of atomic shells?" "Does there not still lurk behind the quark hypothesis the notion, long ago refuted by experiment, that we are able to distinguish simple and composite particles?"

I would now like to take up briefly a few questions of detail. If the SU_3 group plays an important part in the structure of the particle spectrum, and this we must assume on the basis of the experiments, then it is important to decide whether we are dealing with a fundamental symmetry of the underlying natural law, or with a dynamic symmetry, which from the outset can only have approximate validity. If this decision is left unclear, then all further assumptions about the dynamics underlying the spectrum also remain unclear, and then we can no longer understand anything. In the higher symmetrics, such as SU_4, SU_6, SU_{12}, $SU_2 \times SU_2$ and so on, we are very probably dealing with dynamic symmetries, which can be of use in the phenomenology; but their heuristic value could be compared, in my view, with that of the cycles and epicycles in Ptolemaic astronomy. They permit only very indirect back-inferences to the structure of the underlying natural law.

Finally, a word more about the most important experimental findings of recent years. Bosons of relatively high mass, in the region of 3–4 GeV, and of long lifetime, have

WHAT IS AN ELEMENTARY PARTICLE? / 87

lately been discovered. Such states are basically quite to be expected, as Dürr in particular has emphasized. Whether, owing to the peculiarity of their long lifetime, they can be regarded to some degree as composed of other already known long-lived particles, is, of course, a difficult dynamical question, in which the whole complexity of many-particle physics becomes operative. To me, however, it would appear a quite needless speculation, to attempt the introduction of further new particles *ad hoc*, of which the objects in question are to consist. For this would again be that misstatement of the question, which makes no contribution to understanding of the spectrum.

Again, in the storage-rings at Geneva, and in the Batavia machine, the total action cross-sections for proton-proton collisions at very high energies have been measured. It has turned out that the cross-sections increase as the square of the logarithm of the energy, an effect already long ago surmised, in theory, for the asymptotic region. These results, which have also been found, meanwhile, in the collision of other particles, make it probable, therefore, that in the big accelerators the asymptotic region has already been reached, and hence that there, too, we no longer have any surprises to expect.

Quite generally, in new experiments, we should not hope for a *deus ex machina* that will suddenly make the spectrum of particles intelligible. For the experiments of the last fifty years already give a qualitatively quite satisfying, noncontradictory and closed answer to the question "What is an elementary particle?" Much as in quantum chemistry, the quantitative details can be clarified, not suddenly, but only by much physical and mathematical precision-work over the years.

Hence I can conclude with an optimistic look ahead to developments in particle physics which seem to me to give promise of success. New experimental findings are always valuable, of course, even when at first they merely enlarge the tabulated record; but they are especially interesting when they answer critical questions of theory. In theory, we shall have to endeavor, without any semi-philosophical preconceptions, to make precise assumptions concerning the underlying dynamics of matter. This must be taken with complete seriousness, and we should not, therefore, be content with vague hypotheses, in which most things are left obscure. For the particle spectrum can be understood only if we know the underlying dynamics of matter; it is the dynamics that count. All else would be merely a sort of word-painting based on the tabulated record, and in that case the record itself would doubtless be more informative than the word-painting.

The Role of Elementary Particle Physics in the Present Development of Science*

There seems to be general agreement that elementary particle physics plays a very important role in contemporary science. This can be seen from the great number of physicists who engage in research in this field and from the enormous budgets of national or international institutions providing the experimental facilities for such research. Therefore I would like, in this chapter, to study this role more closely; to investigate in detail the relations of this field to other parts of physics or other sciences; to discuss the reasons and the implications of this activity as part of what one calls big science; and, finally, to look at the results obtained in the past decades and to draw the consequences for the conceptual framework of science.

When, in the beginning of the century, the work of Rutherford and Bohr gave a first insight into the structure of the atom, it was easily seen that an understanding of the elec-

---

*Lecture to the Academy of Sciences in Stockholm on April 24, 1974. Published in the *Documenta* of the Stockholm Academy, 1974. (Written in English.)

tronic shells of the atom would have the most important consequences for many fields of physics or science in general. The behavior of matter in its different forms, as solid body, liquid or gas, should be explained on this basis; special properties like crystal structure, electric conductivity, elasticity of solid bodies, superfluidity of liquids should be understood, and the color of gases in discharge tubes would be derived from the electronic orbits in the atom. The extensive material collected by the chemists about molecules, the properties of chemical compounds, the mechanism of their reaction, could be analyzed and interpreted by this new knowledge of the atom. There could be no doubt that many important practical applications would follow and, actually, after 1930 when the behavior of the outer parts of the atom had been understood, there was a rich harvest in the physics of solid bodies, in low temperature work and in astrophysics.

A corresponding development was *not* expected in the beginning of nuclear physics. In most phenomena—except radioactivity—the atomic nucleus acts as an unchangeable unit and, even as late as the mid-1930s, one could doubt whether there would ever be a technical application of nuclear physics. But after von Weizsäcker and Bethe had demonstrated that the stars are big atomic furnaces, and after Hahn had discovered uranium fission, the way was open to the enormous technical development that followed.

Can we expect comparable consequences in elementary particle physics? For historical reasons we use the term *elementary particle*, if we use it at all, for objects with baryonic number 1 or 0, charge 1 or 0, leptonic number 1 or 0; it is the behavior of these particles that has been studied extensively during the last twenty years. Again, in the majority of natural phenomena, these particles act as unchangeable units, and it is only by means of the big accelerators that we can change them, transmute them into very unstable objects that finally decay into the few stable objects known as electron, proton, photon, neutrino. There seems to be little room for technical applications. Of course we can combine these objects to get bigger compounds and thereby produce reactions and obtain energy, but these processes belong to nuclear or atomic physics and are already well known.

So we may first ask what relations could be established between a thorough knowledge of elementary particle physics and other branches of our science. The experiments of the last two decades have revealed a rather consistent picture of this world of particles. If, in a big accelerator, two particles undergo a very energetic collision, the result should not be called a division of the colliding particles. Actually, what happens is a production of new, mostly unstable particles out of the kinetic energy of the colliding objects, according to the laws of special relativity. Energy becomes matter by assuming the form of particles. The spectrum of possible particles is as complicated as the spectrum of stationary states of atoms or molecules or nuclei. The particles are characterized by quantum numbers, i.e., by their symmetry, their behavior under the fundamental transformations, as atoms or molecules are. Therefore the field of particle physics can be adequately compared, for example, with the field of chemical reactions in gases. What is required is the knowledge of many different objects, molecules in the one case, particles in the other case, and of their reactions under collisions. During the last twenty years many data about elementary particles have actually been accumulated, and we may ask about the relevance of these data for other parts of science.

Let us start with the theoretical side of the picture. Both the methods and the results can be relevant in other fields. The methods by which we discuss the processes in particle physics are similar to those used for reactions between atoms or molecules or electrons, e.g. in quantum chemistry. These methods belong to a theoretical field called "many body physics," and any progress which is achieved in particle physics can be useful in this field, e.g., in the study of chemical reactions or in investigations about the behavior of excited states in solid bodies. And, conversely, these excited states—named polarons, excitons, etc.—are perhaps the best image found in nonrelativistic field theory. So there has been a fruitful exchange of ideas between both fields.

With regard to the results, the most important application of our knowledge in particle physics would probably be in nuclear physics. The forces between the constituent protons and neutrons in a nucleus are not very well known. There are phenomenological descriptions of these forces which lead to a fair agreement in calculations concerning the stationary states of nuclei; but already the determination of three-body forces is very uncertain. The forces are, at least to a large extent, produced by the exchange of particles, primarily by bosons, such as pions, ρ-mesons, etc.; therefore a knowledge of particles and their interactions contributes to the knowledge of the nuclear forces.

In this connection one could imagine a future influence of particle physics on astrophysics. Recent observations have revealed the existence of stars of extreme density, higher by a factor of the order 10^{15} than the density of normal matter. These stars, pulsars or neutron stars, seem to consist of neutral nuclear matter. The interactions responsible for the physical properties of this matter should be similar to the interactions in atomic nuclei, except for gravitation which plays a decisive role in the stars, but not in nuclei. For stars with a mass considerably larger than the mass of the sun one would expect a further contraction on account of gravitation, and in this case nuclear physics would give no information about the inner structure of matter of such density. However, when the natural laws responsible for the existence of the elementary particles are completely known, this knowledge could perhaps supply sufficient information about the conditions in the interior of such astrophysical objects, including the famous black holes.

So there are interesting applications of particle physics in solid state physics, nuclear physics and astrophysics, but they are scarcely sufficient to explain the great interest in and enormous effort on behalf of particle physics. Actually, at this point the question is raised whether the natural laws from which the particles are constructed could act as a general basis for all branches of physics; whether the law is fundamental in this sense. This was in fact the intention of atomic physics or atomic philosophy from its very beginning: to find fundamental laws from which nature could be understood. But I would prefer to come back to this side of the problem in the last part of this chapter and first discuss the great experimental and technical activity in elementary particle physics.

From an experimental point of view, particle physics is a natural continuation or extension of atomic physics and nuclear physics. The outer parts of the atom, the electronic shells which were in the center of interest in the early twenties, could be influenced by small forces. Electric and magnetic fields could produce changes visible in the spectra of the atoms, electrons accelerated in discharge tubes of a few volts could put the atom in an excited state, and the light emitted gave valuable information about the dynamical structure of the shells. Hence experimentation in those days could be carried out in small laboratories by means of equipment which was extremely cheap compared with the tools of research in modern institutes. But the atomic nuclei could not be investigated by this kind of equipment, for lifting a nucleus to an excited state requires an energy which is roughly a million times greater than that needed for exciting electronic shells.

So Cockcroft and Walton constructed a high tension apparatus, the cascade generator, Lawrence built the cyclotron and, by means of protons which had been accelerated by a voltage of the order one million volts, the atomic nuclei could be excited, transmuted into other nuclei, and very many new, unstable radioactive nuclei could be formed. At the same time new instruments for the registration of nuclear fragments were developed; counters of various types, cloud chambers which could be triggered by the events and new techniques for measuring coincidences were invented. In this way nuclear physics became an important field of science during the thirties, even before Otto Hahn discovered the fission process of uranium and thereby opened up the way for an enormous technical development. The practical applications and their political consequences could at least be considered a sufficient justification for the rather high budgets needed for nuclear research. During the Second World War the United States' budget for nuclear research ran into billions of dollars and the whole attitude of the community toward technology was radically changed. But there was a third step which could already be envisaged by the physicists before the war.

In cosmic radiation, particles were occasionally found with an energy still a thousand or a million times larger than the energies needed for nuclear transmutation. A collision of such particles could possibly lead to a transmutation or a splitting of even those objects which had been considered elementary particles—final unchangeable units of matter, namely the proton or electron. Theoretical arguments suggested that, in very energetic collisions between two "elementary" particles, new particles can be created, possibly many particles, and that such processes should not be named division or splitting or excitation of particles, but rather that one should simply speak about the transmutation of energy into matter. The experiments on cosmic radiation before the war gave slight indications in this direction, but no indisputable proof for the multiple production of particles.

In the years after the war, therefore, physicists discussed the possibility of constructing huge accelerators in order to make a systematic study of these processes. The realization of such plans would require a technical effort and a budget far beyond anything that had been considered for fundamental research before. But, on the one hand, the American government had been accustomed to allot very large sums to nuclear research during the war while, on the other, the problem of the smallest units of matter seemed so important and so exciting that efforts in this direction became feasible. You know that many big accelerators, growing in size, have been constructed since the war. First there was the cosmotron in Brookhaven with 3 GeV energy of the accelerated protons, then the Bevatron in Berkeley with 6, the CERN machine and a corresponding accelerator in Brookhaven with 30 GeV; the Russian machines in Dubna with 10 and in Serpuchov with 70 GeV followed. At present the highest energies are obtained in the Batavia accelerator and in the storage rings in Geneva. I do not want to describe the almost incredible precision and reliability necessary for getting such machines to work, the enormous technical skill which has been demonstrated in their construction. One can only admire what has been done by engineers and physicists in fulfilling these tasks.

During this development elementary particle physics has become a part of what one calls big science; and since everything in this world has its price, the style of experimental physics has had to undergo serious changes during this expansion. I must confess that I have always felt a little uneasy about these changes and, therefore, I would like to discuss them in more detail. The difficulty is not that the cost of a big accelerator runs into billions of dollars and that its construction takes many years. This is not unusual for a big engineering project and has little to do with fundamental research in physics. The real difficulty is that a single experiment with such an accelerator needs a long planning stage, a large budget and many years and many people for its completion.

Let us speak for a moment about the old times. Niels

Bohr told me about one of the early experiments of Lord Rutherford at McGill University in Montreal. In studying the behavior of radioactive substances Lord Rutherford had one day come to the conclusion that in the decay of radium an inert gas is formed, later called radon or radium emanation, and that it should be possible to liquefy it at very low temperature and thereby to concentrate it and investigate its properties. Rutherford ordered low temperature equipment from Europe and, after it had arrived by boat in Montreal, no member of the institute was allowed to leave the building before the experiment was carried out. Everyone had to be busy in mounting the equipment, preparing the counters, etc.; the assistants had to stay all night and, actually, after thirty-six hours, the liquefaction was successful and demonstrated the existence of radon. You can imagine the satisfaction of Rutherford to see that his idea had been correct, and that of the assistants who had been able to actually demonstrate it.

Now take a comparable situation in modern elementary particle physics. Seven years ago a good, young physicist from our institute found out that the equipment of one of our CERN teams should probably be sufficient to measure an interesting quantity, the coupling of η-mesons to nucleons. This quantity was important because different theoretical pictures gave very different estimates, hence the quantity would be relevant in the theoretical interpretation. The memorandum of the young physicist, written in October 1967, was accepted as sensible; he entered the team, and the hope was expressed that the experiment could be carried out within one year. But of course first another experiment, that had been started before, should be finished.

This other experiment took much longer than expected. The equipment had to be improved and new counters had to be installed before one could really be satisfied with the reliability of the results; several years went by before the experiment was completed. The experience gained in these years taught the team that the new experiment suggested by the young physicist would require another improvement of the counters, therefore it might involve too high a risk to start this experiment at once. It might be better first to continue the line of the older experiment and thereby test the new equipment; this plan would also fit in better with the general problem of CERN. This extension of the older experiment took several years; the proposal for the new experiment was finally given to the CERN committee last year; there is definite hope that a positive decision will be reached this year and that the experiment can be carried out next year. Such a time span of eight years—between the first notion of the experiment and its completion—may be above the average; but six years would perhaps be normal with a big accelerator. In any case it is obvious that for a young and active scientist the time-scale of this kind of research raises a serious problem.

A second difficulty is introduced by the necessity for specialization. The equipment needed for an experiment with a big machine consists of many parts, each of which requires a specialist for construction and operation. Therefore the physicist who is in charge of such a part is fully engaged with his apparatus and has little time to think about the experiment and its purpose, let alone particle physics.

And, finally, the decision to carry out a project cannot be made by a single physicist. There are always several teams applying for time at one of the beams of the accelerator and, hence, committees have to decide priorities. But the ideas come from the single physicists; committees have no ideas. A responsible committee must always tend to continue along the line that has been successful in the past and to avoid the unusual and the risk connected with new ideas.

All this is absolutely necessary; it is the unavoidable consequence of the fact that elementary particle physics has become a part of "big science." We must accept it. But we should keep it in mind when, for example, as advisors to our governments, we discuss budgets for the various parts of science. The organization for scientific research is an important task for any modern government, and very often senior physicists have to advise the administration on priorities. The budget for constructing and running a big accelerator is comparable with the budget for founding and running a new university. Therefore, it is obvious that many arguments, partly political, partly scientific, must be weighed before decisions can be reached. I would like to discuss a few of these arguments.

There is, first of all, the international character of particle physics. In hardly any other field of science has international collaboration been so necessary and so successful as here. The absence of immediate technical application protects this field from interference by economic or national interests. The very high costs for an accelerator laboratory can therefore be easily divided among many governments, and the scientific life in such an institution contributes quite substantially to the understanding between the physicists, the engineers and the officials of the various nations. The CERN organization in Geneva seems to me the most efficient international establishment of this kind.

At the same time, such an international laboratory may well stimulate technical programs in special fields, and this stimulus may have its effect in any of the cooperating nations. Good experiments in particle physics require the most modern technology, and may therefore sometimes help in developing this technology. But this use, the technical "fallout" of the big accelerators, should not be overemphasized. Any branch of "big science" will produce similar effects, perhaps in different fields; we do not gain in this way an argument for giving a high priority to particle physics.

The tendency for concentrating particle physics in big international centers is liable to limitation, however, by the time-scale of the experiments. The cooperation between the universities and research institutions of the member states on the one hand, and the international center on the other, would require that frequently physicists or teams from one of the national institutions should go to the center for one or two years, carry out their experiment and then go back again, thereby transferring the experiences and the scientific and technical knowledge from the center to the national institution. But when a normal experiment takes six years or more the situation is quite different. A physicist who, for six years, has lived with his family in the neighborhood of the international center will not, as a rule, want to go back. He feels at home in the new surroundings; the children go to school there; the facilities for scientific work are better in the center than in his old university. So, frequently, the modern scientific life of the center will not be brought back to the member state. On the contrary, the member state may lose some of its most gifted young scientists to the international center. This difficulty can probably be met only by a considerable activity in particle physics in the member state itself. If a national institution, perhaps with some smaller specialized accelerator, stimulates interest in particle physics, then a flow of information will go back and forth between the national institution and the international center, and the member state gets full use from its contributions. But his national institution again requires a large budget.

So, finally, we are left with the question of whether particle physics is such a fundamental field of science that great material sacrifices for its exploration can be justified. The fundamental character of particle physics will be the object of the last part of this chapter; for the moment I will only discuss one partial and practical aspect of this problem, namely the question: Do we get more fundamental, more essential information when we go to higher and higher energies of the colliding particles, i.e., when we build bigger and bigger accelerators? It is an almost generally accepted doctrine that this must be so. Each time in the past the step to higher energies has opened up a new field, so why should it not go on in the same way in the future? In the last part of this chapter I will give a few reasons for suspecting that this doctrine may be wrong. But even if the doctrine is accepted, the economic and social problems in the present world will make it extremely difficult to finance the construction of accelerators much bigger than those under construction now. Therefore we expect that, during the next decade, particle physics will have to rely entirely on the existing machines and those under construction now; the storage rings in Geneva will produce the processes with the highest energy. If new experimental results should attract interest to still higher energies, one could perhaps use cosmic radiation for a preliminary exploration, as was done in the early fifties. Such experiments would probably be cheaper than the construction of bigger accelerators, but the results can not be equally reliable.

Some physicists plead for bigger accelerators in spite of the economic, political and social implications, and they compare these machines of our present world with the Egyptian pyramids or the cathedrals of medieval times. They argue that those huge monuments were erected as symbols for the core of society of humanity's relation to the highest power. The interpretation of the world that was basic to that society was made visible by the great symbol. And, in the same way, the big accelerators of our time could be the symbols for our scientific interpretation of the world.

I do not know whether I can quite approve of this kind of argument. It is true that science seems to be the center of confidence in our time. In medicine, in agriculture, in technical applications, we rely on the correctness of science. At the same time we feel that this interpretation of the world is too narrow. It leaves out essential parts which have belonged to the substance of the older religions. It leaves them out so completely that it is even difficult to speak about them. But the unrest of the young generation, and many other signs of unsteadiness, seem to indicate that there is a gap that must be filled. I can therefore scarcely believe that the common man on the street will take a big accelerator, which looks like a factory from the outside, as a symbol for his interpretation of the world; but I may be wrong.

Perhaps we should leave this question unanswered and ask, rather, in what sense elementary particle physics can be considered fundamental in science. It was certainly the intention of atomic physics from the beginning to penetrate through the visible phenomena to fundamental structures, to an understanding of nature. The historical way of science has led from chemistry to the Bohr-Rutherford model of the atom, from this picture to the hypothesis that atomic nuclei consist of protons and neutrons and to the idea that all matter is composed of the three elementary particles—proton, neutron, electron—and finally to a spectrum of particles which can be created when energy is transmuted into matter. There have been surprises on this way which have caused essential changes in the conceptual framework of physics, and these changes are important in discussing the question of whether we have reached fundamental structures in particle physics, and what these structures are.

The first surprise concerned the limitations in the use of Newtonian mechanics. In the time before Planck had discovered the quantum of action, mechanical processes were discussed with the concepts of classical mechanics, and they had always been understood that way. The stability of the atom, however, could not be interpreted by this theoretical scheme. When an atom has been perturbed by outer forces in a chemical reaction, by collisions in a discharge, by electro-magnetic fields, it always returns finally to the same normal state. A planetary system of electrons moving around a nucleus would not show this kind of behavior. This fact was the starting point for Bohr's hypothesis of the discrete stationary state, which introduced Planck's quantum of action into mechanics. After quantum mechanics had been formulated, the word *state* meant something different from what it had meant in classical mechanics. In earlier physics the reaction of a system to outer forces, e.g., in the process of observation, was uniquely defined by what one called the state of the system. In quantum mechanics, only probabilities for the reaction could be defined when the state was known. The change of the state in the course of time was given by a dynamical law, as in Newton's mechanics. Under constant outer conditions, some states do not change in time and are therefore called "stationary states"; these states belong to discrete energy values and are determined mathematically as eigen-solutions of some system of linear equations. By this concept of the discrete stationary state the old idea of the atom had undergone a very essential change.

In the old philosophy, the atom was an unchangeable fundamental unit of matter. Bohr's atom, however, as a discrete stationary state of a mechanical system, was not unchangeable. It could be perturbed by outer forces, in collisions, in chemical reactions, but it would be restored after the perturbation had ceased. Atoms are perturbed and restored again and again during the interaction of matter. This characteristic behavior of the discrete stationary state is connected, in the mathematical description, with its behavior under certain symmetry operations. If the dynamical law determining the system is invariant under an operation, e.g. under rotation in space, the mathematical representation of the discrete stationary state will also be a representation of the group of rotations, and will therefore not change in time under this dynamical law. This connection between the discrete stationary states and the symmetries of the system was, of course, not so clear when Bohr formulated his hypothesis, but it was found later as the result of thorough studies of quantum mechanics by Wigner and the mathematicians.

When this feature of the discrete stationary state had been understood, the first but somewhat rash conclusion was that the chemical atom was not an atom, but a compound system consisting of protons, neutrons and electrons, and that the latter elementary particles were the really unchangeable fundamental units of matter. This could not be quite true for the neutrons, because they decayed into protons, electrons and neutrinos, but protons and electrons could be elementary, and this was the generally accepted view for some time.

The next surprise came through Dirac's theory of the electron and the discovery of the positron. In a relativistic theory, the square of the energy is connected with the square of the momentum, and this quadratic relationship produces a doubling of states; the electron is supplemented by the positron. Therefore, electron-positron pairs can be produced by radiation. Later it was also found that electron-neutrino pairs could be created in radioactive decay. This result indicated that even electrons are not unchangeable units of matter; they can be created or annihilated. Energy can be transmuted into matter by taking the form of particles. This general statement could soon be confirmed by investigating very energetic collisions between particles. We know now, from many experiments in the cosmic radiation or by means of the big accelerators, that as a rule from such collisions many particles emerge, most of them unstable like pions, kaons, and hyperons. And these processes should not be considered as divisions or disintegrations of the particles; they demonstrate the transmutation of energy into matter. Besides the stable particles, proton and electron, the experiments have revealed a very complicated spectrum of unstable particles, which seem to have all the properties of the discrete stationary states in Bohr's theory. They can be characterized by quantum numbers—and that means by symmetries—and just as an excited state of the hydrogen atom can decay into a photon and a normal state of hydrogen, so the pion can decay into muon and neutrino. So, finally, the concept of an unchangeable unit of matter has disappeared; the process of division has lost its meaning, and every particle can be considered as a compound system, if we want to consider it that way. A proton can be taken as composed of kaon and A-hyperon, an electron as composed of pion and neutrino; they are no more elementary than a hydrogen atom.

With this knowledge in mind let us come back to our original question: Has particle physics led us to fundamental structures, to a real understanding of nature? It has certainly not led to fundamental particles. The spectrum of particles is as complicated as the table of molecules in chemistry. One can argue that protons and electrons play a more dominant role in our world than the other particles, just as one can say that water molecules are more important than many other molecules. But one can not see any fundamental difference. Another possibility has been discussed; namely, that in later experiments, at still higher energies, some new particles could be discovered, e.g. quark particles of charge 3, which could be more fundamental than the others. But the existence of such particles could not get around the transmutation of matter into energy, and vice versa, hence one does not see in what sense they could be more fundamental.

Perhaps, at this point, we should briefly come back to the question: Should we built bigger and bigger accelerators? We can certainly not exclude the possibility that at higher energies new surprises could occur, e.g. a new type of shower or particle could be found. But we have to add that our present knowledge does not give any indication in this direction. It is quite possible, from the present evidence, that in the storage rings in Geneva the asymptotic region has been reached; and even in the cosmic radiation, which goes up to something like 10^5 GeV, no unusual processes have been observed. We do get a consistent picture from our present experimental knowledge, and this picture seems to allow a natural theoretical interpretation of the relations between the various forces and particles, between strong, electromagnetic, weak interactions and gravitation.

Again we have to ask whether, in particle physics, fundamental structures of nature have been found. I think one can say that the fundamental symmetries have been revealed. The term *fundamental symmetry* means that the natural law which determines the spectrum of particles and their interactions is invariant under certain groups of transformations. These groups define the total space in which the real world occurs. The most important groups are probably the Lorentz group defining space and time, the group SU_2 related to electromagnetic phenomena and the scale group responsible for the asymptotic behavior at extremely high energies. It is this group structure which we actually investigate in particle physics, and this structure is fundamental. Such a radical change in the conceptual framework of science—from the fundamental particles to the fundamental symmetries—is not easily accepted; it is difficult to get away from questions like: Of what does matter ultimately consist? Can protons be divided by very energetic collisions? But I think it is a final result of the experiments that such questions have no meaning. The quest for the fundamental symmetries is, however, a sensible problem, even if it looks very abstract. Final answers can be given only by investigating many details in the phenomena, both experimentally and theoretically; and this is what actually happens in the big laboratories of particle physics.

Regarding the role of this branch of physics in modern science, the conclusion seems to be that particle physics does actually inform us about fundamental structures of nature. These structures are much more abstract than we had hoped for fifty years ago, but they can be understood. The great effort of our time in this field can be taken as an expression of the human endeavor to penetrate to the innermost center of things. I cannot regret that this center is not material, that it has to do rather with the ideas than with their material image. In any case we should try to understand it.

Encounters and Conversations with Albert Einstein*

The city of Ulm, where Albert Einstein was born, and the Einstein Building of the Ulm public high school, are certainly suitable places in which to tell of encounters and conversations with Einstein. The word *encounters* here must be taken to refer, not only to personal meetings, but also to encounters with Einstein's work, and even quite early on such encounters played a part in my life.

Let me begin, therefore, with the earliest event of this kind that I am able to remember. I was fifteen years old at the time, a student at the Max-Gymnasium in Munich, and had a great interest in mathematical questions. There came into my hands one day a slim volume, containing a collection of scientific articles, in which Einstein had presented in popular form his special theory of relativity. I had met the name Einstein occasionally in the newspapers, and had also heard of the theory of relativity, of which I had gathered that it was extraordinarily hard to understand. I

---

*Unpublished manuscript of a lecture delivered at the Einstein-Haus in Ulm on June 27, 1974.

found this a special attraction, of course, and so attempted to make a most thorough study of this small work. After some time I thought I had quite understood the mathematics—at bottom, after all, it involves only a particularly simple case of the Lorentz transformation—but I soon realized that the real difficulties of this theory lay elsewhere. I was called upon to recognize that the concept of simultaneity is problematic, and that in the end the question whether two events at different places are simultaneous depends on the state of motion of the observer. I found it extraordinarily difficult to think my way into this set of problems, and even the fact that Einstein had seasoned his text on occasion with such interjections as "dear reader" in no way made understanding easier. I was left, nonetheless, with a clear sense of what Einstein was after, a realization that his claims quite plainly involved no internal contradictions, and lastly, of course, a burning desire to penetrate more deeply into relativity theory at a later date. I therefore resolved, during my subsequent studies at the university, at any rate to attend lectures on Einstein's theory of relativity.

Thus my original wish to study mathematics was imperceptibly diverted toward theoretical physics, of which at that time I still knew very little. But I had the great good fortune, at the outset of my studies, to encounter an outstanding teacher, Arnold Sommerfeld of Munich, and the fact that he was an enthusiastic exponent of relativity theory, and also had close personal contacts with Einstein, provided the best of auspices for my initiation, in every detail, into this new field of science. It was a not infrequent occurrence for Sommerfeld to read out to his seminar letters that he had just previously received from Einstein, and for the whole class to be then exhorted to understand and interpret Einstein's text. Even today, I still recall such discussions with great pleasure, and from Sommerfeld's observations I eventually came to feel that I already knew Einstein a little, too, in an almost personal way, although I had never yet seen him. Now before recounting my initial, albeit abortive, attempt to become personally acquainted with Einstein, I must first say something of another field of science which at that time held me in thrall, and in which the name of Einstein also plays an important part.

The central interest of my teacher Sommerfeld, even in his own research, was atomic theory, and more precisely that application of quantum theory and the atom concept whereby Niels Bohr, in 1913, had taken the decisive step into modern atomic physics. I was attending Sommerfeld's lectures and seminars on this subject from my earliest days as a student, although I had certainly not yet acquired the qualifications for doing so. But the fascination emanating from Sommerfeld's passionate interest in these questions made up for many a disappointment that arose when the effort to understand proved fruitless. In this connection, there was much talk of Einstein's hypothesis of light-quanta, and I must explain what that was about. In Sommerfeld's lecture course we first learned the traditional view, which had been generally accepted since Maxwell's day, that light can be interpreted as an electromagnetic wave motion, differing only in its wave-length from radio waves on the one hand, and Röntgen rays on the other. In contrast to this, Einstein, in keeping with Planck's quantum theory, and on the basis of particular experiments on the photoelectric effect, had put forward the hypothesis that light consists of very small energy-quanta, and that a light-ray can therefore be compared to a hail of many tiny pellets. These two views were so radically different that I could make nothing at all of Sommerfeld's assurance that both ideas seemed to possess a certain element of truth. Einstein had again come up with a claim which called in question all the basic assertions of earlier physics; but this time there was also no proof that the new viewpoint did not lead to internal contradictions. On the contrary, the interference phenomena so frequently observed and studied seemed to stand in unbridgeable conflict with the hypothesis of light-quanta. But in atomic physics there were other such insoluble contradictions as well. According to Bohr, the atom consisted of a relatively heavy atomic nucleus, surrounded by electrons, just as the sun is girdled with planets. To this planetary system the same mechanical laws were applied as those used in astronomy, namely, the laws of Newtonian mechanics. At the same time, however, it was claimed that there could only be quite specific electron pathways, marked out by quantum conditions. This statement contradicted Newtonian mechanics, since according to the latter, an external perturbation could easily convert a quantum orbit into one that was not permissible in quantum theory. But in reality, it seemed, for example, that an incoming light beam would lift the electron discontinuously from one quantum orbit to another. At this point, too, along came Einstein with his hypothesis of light-quanta; he held the process of light emission or absorption to be a statistical one, in which light-quanta are ejected or admitted by the atom with a certain frequency. The frequencies for such processes were determined by so-called transition probabilities, and Einstein, in a celebrated memoir of 1918, had succeeded in deriving from this notion Planck's law of thermal radiation.

In the early years of my studies, therefore, when I was trying to penetrate deeper into what was then modern physics, I kept on running into Einstein's name and work, and the wish to be personally acquainted with the author of so many new ideas kept growing from year to year. But my first attempt to see this wish fulfilled proved a failure. It was in the summer of 1922. The Society of German Scientists and Physicians had announced that, at the congress to be held in Leipzig, Einstein was to give one of the main addresses, and this on the general theory of relativity. Sommerfeld suggested to me that I should visit this session and attend Einstein's lecture; he would then introduce me to Einstein in person. But it was a time of political unrest. The bitterness at Germany's defeat in World War I, and at the harsh conditions imposed by the victors, had not yet died away, and disagreement about what was to be done repeatedly led to severe civil disturbances. At this time, too, there appeared the first symptoms of anti-Semitism, which were stirred up by right-wing radical groups.

In the summer of 1922, shortly before this scientific congress in Leipzig, the then foreign minister, Walther Rathenau, was murdered by nationalist terrorists. It was a deliberate attempt to prevent any effort at a settlement. Political passions again flared high, and the anti-Semitic movement began to direct its vengeance upon Einstein too, since he was a Jew, and enjoyed an especially high esteem in the learned world of Germany. So just before the Leipzig session it was decided, at Einstein's request, that he himself should not speak there, but that von Laue should take over his lecture. I did not know this when I went to Leipzig, and merely wondered at the ominous political excitement that was to be sensed among most of those attending the session. When I sought to enter the great assembly hall, in order to listen to Einstein's lecture, a young man thrust into my hand a red leaflet, reading more or less to the effect that the theory of relativity was a totally unproved Jewish speculation, and that it had been undeservedly played up only through the puffery of Jewish newspapers on behalf of Einstein, a fellow-member of their race. I thought at first that this was the work of one of those lunatics, who do, of course, occasionally frequent such meetings. But when I found that the red leaflet was being distributed by the students of one of the most respected of German experimental physicists, obviously with his approval, one of my dearest hopes disintegrated. So science, too, could be poisoned by political passions; so even here it was not always a question solely of truth. I became so agitated that I could no longer really take in the lecture. I was sitting in the hall a long way from the rostrum, and quite failed to observe that von Laue was speaking in place of Einstein. Even after the meeting I made no attempt to seek out Einstein's acquaintance, but boarded the first train back to Munich. My first personal encounter with Einstein did not occur for another four years, during which great and incisive changes took place in physics.

Of these changes, a brief word must now be said. The contradictions that I mentioned earlier, which had arisen in the quantum theory of atomic structure, had become ever crasser and more insoluble as time went on. New experiments, for example the so-called Compton effect and the Stern-Gerlach effect, had shown that without a radical change in the forming of physical concepts such phenomena can no longer be described. In this situation I thought of an idea that I had read in Einstein's work, namely the requirement that a physical theory should contain only quantities that can be directly observed. This requirement, so the idea went, provided a guarantee of connections between the mathematical formulae and the phenomena. The following-out of this notion led to a mathematical formalism which really seemed to fit the atomistic phenomena. In conjunction with Born, Jordan and Dirac, it was then elaborated into a closed quantum mechanics, and appeared so convincing that there really could be no further doubt of its correctness. But we still did not know how this quantum mechanics should be interpreted, how we should talk about its content.

At this time, in early 1926, I was invited by the Berlin physicists to speak at the colloquium there on the new quantum mechanics. Berlin was then the citadel of physics in Germany. Here Planck, von Laue, Nernst and above all Einstein were teaching. Here Planck had discovered the quantum theory, and Rubens had confirmed it by his measurements of thermal radiation. And here Einstein, in 1916, had formulated the general theory of relativity and the theory of gravitation. Einstein would thus be in the audience; I would make his personal acquaintance. It goes without saying, that I now prepared my lecture with the greatest care. For I wanted, in any event, to make myself intelligible, and more especially to get Einstein interested in the new possibilities. The lecture went off more or less as desired; there were good and helpful questions asked in the discussion that followed. I saw that I had secured Einstein's interest, when immediately afterwards he invited me to come home with him, so that there we might discuss the problems of quantum theory more thoroughly and without distraction.

For the first time, therefore, I now had an opportunity to talk with Einstein himself. On the way home, he questioned me about my background, my studies with Sommerfeld. But on arrival he at once began with a central question about the philosophical foundation of the new quantum mechanics. He pointed out to me that in my mathematical description the notion of "electron path" did not occur at all, but that in a cloud-chamber the track of the electron can of course be observed directly. It seemed to him absurd to claim that there was indeed an electron path in the cloud-chamber, but none in the interior of the atom. The notion of a path could not be dependent, after all, on the size of the space in which the electron's movements were occurring. I defended myself to begin with by justifying in detail the necessity for abandoning the path concept within the interior of the atom. I pointed out that we cannot, in fact, observe such a path; what we actually record are frequencies of the light radiated by the atom, intensities and transition-probabilities, but no actual path. And since it is but rational to introduce into a theory only such quantities as can be directly observed, the concept of electron paths ought not, in fact, to figure in the theory.

To my astonishment, Einstein was not at all satisfied with this argument. He thought that every theory in fact contains unobservable quantities. The principle of employing only observable quantities simply cannot be consistently carried out. And when I objected that in this I had merely been applying the type of philosophy that he, too, had made the basis of his special theory of relativity, he answered simply: "Perhaps I did use such philosophy earlier, and also wrote it, but it is nonsense all the same." Thus Einstein had meanwhile revised his philosophical position on this point. He pointed out to me that the very concept of observation was itself already problematic. Every observation, so he argued, presupposes that there is an unambiguous connection known to us, between the phenomenon to be observed and the sensation which eventually penetrates into our consciousness. But we can only be sure of this connection, if we know the natural laws by which it is determined. If, however, as is obviously the case in modern atomic physics, these laws have to be called in question, then even the concept of "observation" loses its clear meaning. In that case it is theory which first determines what can be observed. These considerations were quite new to me, and made a deep impression on me at the time; they also played an important part later in my own work, and have proved extraordinarily fruitful in the development of the new physics.

Our conversation now turned to the special question of what happens in the passage of the electron from one stationary state to another. The electron might suddenly and discontinuously leap from one quantum orbit to the other, emitting a light-quantum as it does so, or it might, like a radio transmitter, beam out a wave-motion in continuous fashion. In the first case there is no accounting for the interference phenomena that have often enough been observed; in the second, we cannot explain the fact of sharp line-frequencies. In reply to Einstein's question I fell back here upon Bohr's position, that we are, of course, dealing with phenomena that lie far beyond the realm of everyday experience, and so cannot expect these phenomena to be describable in terms of the traditional concepts. But Einstein was not altogether happy with this excuse; he wanted to know in what quantum state, then, the continuous emission of a wave was supposed to take place. I then produced the comparison with a film, in which the transition from one picture to another often does not occur suddenly, the first picture becoming slowly weaker, the second slowly stronger, so that in an intermediate state we do not know which picture is intended. Thus in the atom also, a situation could arise in which for a time we do not know which quantum state the electron is in. But with this interpretation Einstein was far less happy still. It could not possibly be a matter of our knowledge of the atom, since it could perfectly well happen that two different physicists know something different about the atom, even though one and the same atom is in question. Einstein scented at once, no doubt, that in this way we were approaching an interpretation in which the statistical character of natural laws is in principle acknowledged. For in statistics it is actually a matter of our incomplete knowledge of a system. But he wanted nothing to do with this, although he himself, in his paper of 1918, had introduced such statistical concepts. He was not willing, however, to grant them any intrinsic significance. I, too, had no idea what to do just then, and we separated in the common conviction that a great deal of work still needed to be done before reaching a full understanding of the quantum theory.

Great changes again took place before we met again, in the autumn of 1927, at the Solvay Congress in Brussels. In 1926, on the basis of earlier attempts by de Broglie, Schrödinger had developed his wave mechanics, and proved its mathematical equivalence to quantum mechanics. But his subsequent attempt at simply replacing the electrons by matter waves proved a failure, and there remained the paradox, that electrons can actually be both particles and waves. The spring of 1927 then saw the birth of the so-called uncertainty relations, whereby the transition to a statistical interpretation of quantum theory was finally completed. And hence they now formed the main topic of the discussion in Brussels. As I have already said, Einstein was unwilling to recognize the statistical interpretation; so he repeatedly tried to refute the uncertainty relations. These relations involve the statement that two determinants of a system, which must both be known at once in classical physics, in order to determine the system completely, cannot, in quantum theory, be exactly known at the same moment; and hence that between the uncertainties or inexactitudes of these quantities there are mathematical relations which prevent an exact knowledge of both quantities. Einstein therefore kept on trying, during the congress, to refute the uncertainty relations by means of counter-examples, which he formulated in the shape of thought-experiments. We were all living in the same hotel, and Einstein was in the habit of bringing along to breakfast a proposal of this kind, which then had to be analysed. Einstein, Bohr and I would usually make our way to the congress hall together, so that even on this short walk a beginning could be made on analysing and clarifying the assumptions. In the course of the day, Bohr, Pauli and I would frequently discuss Einstein's proposal, so that already by dinner-time we could prove that his thought-experiments were consistent with the uncertainty relations, and so could not be used to refute them. Einstein admitted this, but next morning brought along to breakfast a new thought-experiment, generally more complicated than the previous one, which was now to effect the refutation. The new proposal fared no better than the old; by dinner time it could be disproved. And so it went on for several days. In the end we—that is, Bohr, Pauli and I—knew that we could now be sure of our ground, and Einstein understood that the new interpretation of quantum mechanics cannot be refuted so simply. But he still stood by his watchword, which he clothed in the words: "God does not play at dice." To which Bohr could only answer: "But still, it cannot be for us to tell God, how he is to run the world."

Three years later, in 1930, there was another Solvay Congress in Brussels, at which the same questions were discussed, and the general outcome was also much the same. Bohr endeavored, with great effort, and careful attention to Einstein's observations, to convince him of the correctness of the new interpretation of quantum theory; but without success. Even the very precise written analysis of Einstein's latest thought-experiments, in which Bohr employed the general theory of relativity for his proof, was unable to persuade Einstein. So there we had to leave it, united in being of different opinions. "We agreed to disagree," as the British say.

I then, unfortunately, did not meet Einstein for many years. For in the meantime the political horizon had again darkened; the National Socialists had come to power in Germany, and to Einstein it was plain that he neither could nor would remain any longer in Germany. He therefore spent a great deal of his time in travel abroad. Many universities all over the world counted themselves lucky if they could secure Einstein as a lecturer, or for a longer visit. The National Socialist revolution of 1933 then wrote *finis* to Einstein's stay in Germany. After various intermediate stops he eventually emigrated to the United States, where he accepted a chair at The Institute for Advanced Study. Here, over the last two decades of his life, he found a lasting refuge, and had leisure there to pursue the philosophical problems which preoccupied him, either in physics or in the field of political controversy. But the unrest of the period did not stop short, of course, even at the perimeter of the Princeton campus, and so in 1939, when the war began, Einstein became involved in political problems of the greatest moment, probably against his real wishes. Hence, in order not to leave the portrait of Einstein all too imcomplete, something should doubtless be said of his attitude to politics, or to public life in general, although I never talked with him on this subject.

At first sight, his position on these general questions appears extremely contradictory. One of his most careful biographers, the Englishman Ronald Clark, writes of him: "Einstein became the great contradiction; the German who detested the Germans; the pacifist who encouraged men to arms and played a significant part in the birth of nuclear weapons; the Zionist who wished to placate the Arabs,"* and who, we have to add, was an emigrant, not to Israel, but to America. We do not, however, wish merely to acquiesce in these contradictions, but must try to discover more exactly the motives that prompted him, in order to come closer to an understanding of his personality.

Einstein had already identified himself as a pacifist early on. He was already supporting the pacifist movement at the beginning of World War I, and in the twenties was still persuaded that nationalism was the main cause of wars. Thus he hoped that with a waning of nationalism the preconditions for a longer lasting peace could be created. He only recognised quite late that even the nascent political movements of the twentieth century, which he partly approved and partly recoiled from, were leading in the event to the formation of great totalitarian power-complexes, which, though no longer national states in the old sense, were nevertheless determined to enforce their claims with a military armament which far surpassed that of the earlier national states. Thus it was only in 1939, with the onset of World War II, that Einstein was really confronted with the problem of pacifism. Even in 1929, he had stated, in reply to a Prague newspaper, that in the event of a new war he would refuse to perform military service. Ten years later he had to ask himself whether this attitude was still justified,

---

*Ronald W. Clark, *Einstein: The Life and Times*. London, New York, 1971, p. 20.

when Hitler and the National Socialists stood upon the other side.

To understand Einstein's answer here, it will be necessary to reflect upon the concept of pacifism. Two attitudes can perhaps be distinguished, which may be designated as extreme or realistic pacifism. The extreme pacifist refuses to participate in military service of any kind, even when the human group to which he belongs, or in which he has decided to live, is most seriously threatened; on such an occasion, he is ready even to surrender himself to destruction, or he tries to flee, till he finds some land upon earth that can offer him asylum. The realistic pacifist makes his decision harder for himself. He believes that in the event of a conflict he should first make an independent judgment of the merits of the case; he knows that these are very differently viewed by the two parties, and he tries, precisely, to see the matter at issue from the other side as well. He knows, moreover, that peace can only be preserved if each of the two sides is ready to make painful concessions. So he tries to persuade his countrymen or fellow believers to abate their own claims, to look less favorably on their own side of the case, and hence to make real sacrifices for the preservation of peace. But if, after all, he concludes upon conscientious examination that the opposition has pitched its claims absurdly high, or that unmitigated evil is here being practised by a human group, he considers it not only his right, but even his duty, to offer resistance to the evil, with weapons if necessary. The difficulty in this second view of pacifism is, that here it is no longer sufficient to be simply in favor of peace. An independent judgment must be made upon the issues, and only then can a decision be reached as to what sacrifices may be required for peace.

In the statements of his earlier period, Einstein was undoubtedly a frequent exponent of extreme pacifism; but—as Clark's biography shows—when war broke out in 1939, he opted in his actions for the second type of pacifism. On the insistent representations of his friends, especially his former Berlin assistant Leo Szilard, he wrote three letters to President Roosevelt, and thereby contributed decisively to setting in motion the atom bomb project in the United States. And he also collaborated actively, on occasion, in the work on this project. He had thus arrived at the conviction, that with Hitler a power so evil had erupted into world history, that it was right and proper to oppose this power, even by the most fearsome means. This was his decision. A French writer once said: "In critical times, the hardest thing is, not to do the right thing, but to know what the right thing is." But at this point I should like once more to drop the question of Einstein's political attitudes, particularly since I never myself discussed such difficult problems with him.

Since I am to tell of my encounters with Einstein, I should not omit to mention a little episode that occurred during the war in the Swabian town of Hechingen. My institute, that is, the Kaiser Wilhelm Institute for Physics in Berlin-Dahlem, was engaged during the war in work on the construction of an atomic reactor. Owing to the increasingly heavy air attacks on Berlin, it had to be evacuated to South Germany in 1943, and found refuge in the little town of Hechingen, in southern Württemberg, in the premises of a textile factory. The staff were billeted here and there among the townsfolk of Hechingen, and chance willed it that I was allotted two rooms in the spacious house of a textile manufacturer. Some weeks later, when I had become better acquainted with the owner, he drew my attention one day to a small house lying diagonally opposite. "Look," said he, "that house belongs to the Einstein family." It was not, indeed, a question of the direct ancestors of the celebrated physicist, but of another branch of the family, who had in fact been living in Swabia for several centuries past. So in spite of his aversion to Germany, Einstein was a regular Swabian. And we may indeed suppose that the uncommon philosophical and artistic activity of this German clan has left its traces, also, in Einstein's thought.

After the war, I met Einstein only on one more occasion, some months before his death. In the fall of 1954, I made a lecture tour in the United States, and Einstein invited me to visit him at his home in Princeton. He was then living in a pleasantly unpretentious one-family house with a small garden on the edge of the Princeton University campus, and the tall trees and park-like approaches to the campus were ablaze on the day of my visit with the vivid reds and yellows of late October. I had been warned beforehand that my visit should last only a short time, since Einstein was obliged to spare himself, on account of a heart condition. Einstein, however, would have none of this, and with coffee and cakes I was made to spend almost the whole afternoon with him. Of politics we said nothing. Einstein's whole interest was focused on the interpretation of quantum theory, which continued to disturb him, just as it had done in Brussels twenty-five years before. I tried to secure Einstein's interest in my view by telling him of my attempts at a unified field-theory, on which he, too, had concentrated the labor of many years. I did not believe, to be sure, that quantum theory could, as Einstein hoped, be regarded as a consequence of field-theory; I thought, on the contrary, that a unified field-theory of matter, and hence of elementary particles, could be constructed only on the basis of quantum theory. The latter, with all its disconcerting paradoxes, was thus the true foundation of modern physics. But Einstein was unwilling to grant so fundamental a role to a statistical theory. He held, indeed, that in the present state of knowledge it is the best account of atomic phenomena, but was not prepared to accept it as the final formulation of these natural laws. The remark "But you cannot believe, surely, that God plays at dice" was several times repeated, almost as a reproach. At bottom, indeed, the difference between the two viewpoints lay somewhat deeper. In his earlier physics, Einstein could always set out from the idea of an objective world subsisting in space and time, which we, as physicists, observe only from the outside, as it were. The laws of nature determine its course. In quantum theory this idealization was no longer possible. Here the laws of nature were dealing with temporal change of the possible and the probable. But the decisions leading from the possible to the actual can be registered only in statistical fashion, and are no longer predictable. With this the conception of reality in classical physics is basically undermined, and Einstein could no longer adjust himself to so radical a change. In the twenty-five years that had passed since the Solvay Congresses in Brussels, the two standpoints had not, therefore, come together, and even on parting we were thinking of the future development of physics with very different expectations. But Einstein was ready to accept this situation without any bitterness. He knew what enormous changes in science he had brought about in his own lifetime, and he also knew how hard it is—in science as in life—to accommodate oneself to changes of that size.

The Correctness-Criteria for Closed Theories in Physics*

In a colloquium held some time ago at the Max Planck Institute for study of the life-conditions of the scientific and technical world, and which dealt with those philosophical foundations of quantum theory to whose understanding von Weizsäcker has contributed so greatly, he raised a question concerning the source of the persuasive power of closed theories in physics; what criteria have entitled us to assume that small improvements in these theories can no longer be undertaken, and that they are therefore in a certain sense final?

Before attempting to answer this question, the concept of a closed theory should be once more briefly explained. By a closed theory we mean a system of axioms, definitions and laws, whereby a large field of phenomena can be described, that is, mathematically represented, in a correct and noncontradictory fashion. The term *noncontradictory* refers here

---

*From *Einheit und Vielheit. Festschrift fur C. F. v. Weizsäcker zum 60. Geburtstag*, ed. E. Scheibe and G. Sussmann, Vandenhoeck & Ruprecht, Göttingen, 1972, pp. 140–44.

to the mathematical consistency and closedness of the formalism to be constructed from the basic assumptions; the term *correct*, to the empirical; it means that experiment must confirm the predictions derived from the formalism. In this sense Newtonian mechanics, for example, is the prototype of a closed theory. Other examples from more recent times are the statistical theory of heat—above all in the shape given to it by Willard Gibbs—the special theory of relativity (including electrodynamics), and finally, quantum and wave mechanics, especially in the mathematical axiomatization of them by von Neumann. Each of these theories possesses a limited field of application, which is essentially established by the very concepts employed in the theory. Outside this field the theory cannot reproduce the phenomena, since it is precisely there that its concepts no longer encompass the course of events in nature.

Now whence comes the conviction that such theories are finally correct? Why do we believe that small changes can never improve them? Here we may start with a historical argument; and point out that even the oldest of the closed theories, Newtonian mechanics, has never been improved by small changes. Wherever the concepts of "mass," "force," and "acceleration" can be employed without reservation, then even to this day the law that "mass × acceleration = force" is unrestrictedly valid. If it should be objected to this, that quantum mechanics can surely be regarded as an improvement on Newtonian mechanics, the answer must be that there it was a matter, not of minor improvement, but of a radical reconstruction of the conceptual foundations. The behavior of electrons in the atom, for example, cannot be understood with the intellectual tools of Newtonian mechanics, though it can with the altogether different conceptual apparatus of quantum mechanics.

A second and somewhat stronger argument for the finality of closed theories is their compactness and manifold confirmation by experiment. From relatively few and simple postulates an infinite wealth of solutions has resulted, of which one in particular is selected, depending on the external conditions of the process under review. Experiments have hitherto confirmed the theory in every single case, and a great many experiments have already been carried out. Admittedly, the theory is not yet strictly proved by that; for a contradiction to the theory could always emerge in a later experiment. Karl Popper, indeed, believes that this situation can justify his claim, that a theory can only be falsified, not verified; to which, though, it has to be objected, in von Weizsäcker's opinion, that even in any experiment which appears to contradict the theory, there are presuppositions involved which have, maybe wrongly, been taken for granted as certain; so that in reality it is not the theory at all, but one of these presuppositions, that gets falsified. The decision as to the correctness of a theory is thus a historical process, stretching over long periods of time; though lacking the probative force of a mathematical proof, it certainly has the persuasive power of a historical fact. The closed theory is never, indeed, an exact replica of nature in the field in question; it is, however, an idealization of experiences which are successfully dealt with by means of the theory's conceptual foundations.

The above-mentioned compactness of the theory, and the idealization of reality effected by this might tempt one to conclude that perhaps its mathematical simplicity and beauty, and thus ultimately an aesthetic criterion, have exerted a governing influence on the persuasive power of the closed theory. But this influence should not be overestimated. For on looking more closely at the existing closed theories, the conceptual foundations appear simple, but not the mathematics. Newtonian mechanics, for example, is formally equivalent to a system of coupled, nonlinear, differential equations, which are by no means simple in their mathematical structure; we have only to recall the many-body problems in astronomy. Gibbs' statistical thermodynamics is centered upon the concept of canonical distribution, in which the simple mathematical behavior of the exponential function can be exploited; but beyond that, there can hardly be much talk of mathematical simplicity. In quantum mechanics we can perhaps speak most readily of a simple mathematical structure, since here the whole well-evolved theory of linear transformations lies at its base. But there, too, the problems connected with the Dirac delta-function mark the boundaries of mathematical simplicity. The compactness of closed theories is therefore more a logical and conceptual compactness than a formally mathematical one. This is no doubt the reason why, in the evolutionary history of closed theories, the physico-conceptual clarification has usually preceded a full understanding of the mathematical structure.

The empirical correlate of compactness is the internal connectedness of many experiments, that is, the knowledge that a deviation of experience from theory in *one* experiment would also inevitably result in such deviation in many others. This, however, has been perceived only in modern times; for ancient or medieval thought there was no connection, say, between the apple's fall from a tree and the moon's motion about the earth. Newton was the first to realize that the apple might also be thrown, that hence there should be no basic difference between falling and being thrown, that the apple, too, can be replaced by heavier bodies, and that ultimately, even the moon can be regarded as a thrown body. A modern space-vehicle provides, as it were, the intermediate object (now also realized in practice) between apple and moon. If, then, the inner connections between many phenomena, as expressed in a closed theory, have held good in innumerable experiments, we can no longer doubt that they are formulated "with final correctness"—this statement being based on the above-mentioned restriction, that we are dealing with an idealization proceeding from a particular conceptual scheme.

But all the criteria so far discussed still leave unanswered an important part of the problem posed by von Weizsäcker; the question, that is, of why, at the very first moment of their appearance, and especially for him who first sees them, the correct closed theory possesses an enormous persuasive power, long before the conceptual or even the mathematical foundations are completely clarified, and long before it could be said that many experiments had confirmed it? Thus Newton, for example, was assuredly not yet in possession of a mathematical theory of coupled, nonlinear differential equations, and for empirical data had hardly much more at his disposal than the Galilean laws of falling bodies and the Keplerian laws of planetary motion; all the same, he wrote the *Principia*. At the beginning of this century, the celebrated transformation of Lorentz and Poincaré was discovered and accepted, even before the conceptual revolution of relativity theory had really made it understood, and although there was really only the Michelson-Morley experiment to go on, in the way of empirical facts. It was much the same in quantum and wave mechanics. So whence comes this immediate persuasive power?

In all probability, the decisive precondition here is that the physicists most intimately acquainted with the relevant field of experience have felt very clearly, on the one hand, that the phenomena of this field are closely connected and cannot be understood independently of each other, while this very connection, on the other, resists interpretation within the framework of existing concepts. The attempt, nonetheless, to effect such an interpretation, has repeatedly led these physicists to assumptions that harbor contradictions, or to wholly obscure distinctions of cases, or to an impenetrable tangle of semi-empirical formulae, of which it is really quite evident that they cannot be correct. We may recall, for example, the attempts to restrict Newtonian mechanics by the Bohr-Sommerfeld quantum conditions, the statements of merely qualitative utility deriving from Bohr's correspondence principle, or the complicated formula for the inertial mass of a moving electron, which seemed to follow from the older electrodynamics. If, then, in the intensive search for new conceptual or formal possibilities, the correct proposal for the closed theory emerges, it has from the very outset an enormous power of persuasion, precisely because it cannot at once be refuted. The research worker who is thoroughly familiar with the field in question is probably of the just conviction that he could at once refute any false proposal for the final theory. If the new proposal appears to be a genuine possibility in which the earlier difficulties are avoided, if it does not immediately run into insoluble contradictions, then it has to be the right proposal. For the conceptual systems under consideration undoubtedly form a discrete, not a continuous, manifold. In the initial stage of the theory's development, errors may then continue to creep in, to be later eliminated; but basically there can be no further doubt about the rightness of the approach.

To confirm this claim that a false proposal is easily refutable, let me look back once more to the past and recount an anecdote relating to the Leipzig seminar of 1930–32, whose participants included von Weizsäcker, along with many of today's well-known atomic physicists, and also the mathematician van der Waerden. At the tea party after the seminar, we used also to talk of more general questions, unconnected with the narrow field of atomic physics, and one day the conversation turned to Fermat's celebrated theorem in the theory of numbers; to the claim, that is, that it is impossible to satisfy the equation $a^n + b^n = c^n$ with integers a, b and c, if the equally integral exponent n is larger than 2. I then asked whether it might not happen that some mathematician could announce that he had refuted Fermat's theorem by producing an example that satisfies the equation; but he chooses for his a, b and c, and especially n, numbers so large that nobody could work out the powers in question, and nobody, therefore, can prove that the equation is wrong. Van der Waerden at once expressed vehement dissent, and offered me a bet: I was to think up any such numerical example, and he would then be able to refute the alleged equation in less than seven minutes; if he did not succeed, I would have won the bet—otherwise he would. I then had a week until the next seminar period, and of course took care to set up the example so that it resisted all the simple criteria I knew of, e.g., that the remainders satisfied the equation for all prime numbers up to 13, and so on. But van der Waerden succeeded, nevertheless, in refuting the example inside three-and-a-half minutes, and hence in winning the bet. He had in fact studied the whole problem so thoroughly, that he had far more criteria at his disposal than any physicist.

False proposals for a closed theory concerning a vast network of physical experience will perhaps not always allow of refutation within three-and-a-half minutes; but they will very quickly be recognized as inadequate by anyone really familiar with the field in question. The surprise produced by the right proposal, the discovery that "yes, that could actually be true," thus gives it from the very outset a great power of persuasion.

Thoughts on The Artist's Journey into the Interior*

The spiritual evolution described by Erich Heller in his book, *The Artist's Journey into the Interior*, is visible in many fields—in painting, music, poetry and philosophy—and it is not surprising that even in science there should be a comparable process, which can perhaps be called the scientist's journey into abstraction. The fact that these two branches of cultural development spring from the same root, has already been indirectly pointed out by Goethe, in that he feared both to the same extent and repeatedly gave insistent warning of their dire consequences.

Once the affinity of these two processes has been assumed as given, it is natural for the questions left unanswered about the artist's journey into the interior to be posed anew about that other journey of the scientist, and perhaps for some illumination to be gained by the comparison. The most important question is doubtless this: where

---

*From *Versuche zu Goethe. Festschrift fur Erich Heller zum 65. Geburtstag*, ed. V. Durr and G. v. Molnar, Lothar Stiehm Verlag, Heidelberg, 1976.

does this journey lead? Can the goal be more exactly described than by the words *interior* or *abstraction*? And what will happen once the goal is reached? Where are we then?

Before discussing the scientist's answer on this point, we must realize that this development, or at least its consequences in art and science, were quite early subjected to strong resistance, which often found expression in the opponent's rage and despair, but have not been able to halve the voyage thus embarked on. Goethe's warning has already been cited. Erich Heller also points to the observations, marked by anger and amazement, that Rilke made about the paintings of a Matisse, Picasso or Braque. Many other such adverse judgments upon modern art can assuredly be found. Similar occurrences are known in the field of exact science: the sometimes embittered attacks conducted against Einstein's theory of relativity, and more recently also against the quantum theory, and the theory of elementary particles, as it pursues its way into ever more abstract regions. Here, too, one may find among the assailants extremely prominent representatives of their fields, and it remains all the more astonishing that their criticism has been rewarded with so little success. This is doubtless connected with the fact that, among the critics, strong emotions have been aroused, which have risen, in individual cases, to hatred, personal vilification and the carrying of the struggle into the political arena. If it is true that hatred springs from impotence, we shall have to conclude that even the critics had no genuine alternative to offer to the course of the journey. So this seems to be the core of the problem, that the dangers of the goal alarm us exceedingly, but that we see no possibility of deviating from the journey, toward other goals. It is all the more important to see how great the goal's dangers actually are.

Let us begin with biology. The journey begins approximately at the point where, in the attempt to understand the manifold shapes of plants, there appeared to the mental eye of Goethe the poet, the *Urpflanze*, which embodies and makes immediately visible, in some measure, the principle on which plants are constructed. But his successors were already enquiring into the role of particular organs—leaf, root, blossom and fruit—and into their construction from cells; and finally the road went on to the structure of cells, to the atomic make-up of their components, and to the processes involved in cell-division. At the goal of the journey, the biologists encountered that bundle of information which is chemically inscribed on the double chain of nucleic acid, as on the punched tape of an electronic computer, and which contains the building-plan of the organism. This totality of information, or its various chemical correlates, can be described as a sort of ur-organism. Yet it has to be remembered that at this lower extreme the boundaries between animate and inanimate matter disappear, so that we could also speak merely of a very complicated molecule. For the comparison with processes in art, there are two other conclusions, perhaps, which are of more importance than these special findings. The biological road into the interior, that is, into abstraction, has not been infinitely lengthy; even at the point already referred to, it has found a well-defined natural termination. In the landscape it has led through, there is still, indeed, an endless amount to explore; but the terminus is not thereby called in question. And secondly: at the goal we are confronted with the old Platonic question about the real. Is this bundle of information really the living creature, or is it merely the form of the latter, whereas the chemical molecules that present it constitute the real living structure? The totality of information is to some extent the Platonic Idea of the creature. And we have returned to the age-old problem of whether the Idea is more real than its material realization. At which point the doubt emerges, as to whether, perhaps, it is merely a question of discovering or defining what the term *real* is to mean.

For physical and chemical enquiry, the road into the interior has taken a very similar course. Goethe, as geologist, was still collecting and examining minerals; the period after him took an interest in the chemical composition of crystals; atoms were once more the smallest parts that a chemical element can be split into, without altering its nature. But at this point the road already leads far into the abstract. The atom, so Bohr's theory tells us, consists of a heavy nucleus and the electrons orbiting around it. Yet already we no longer know exactly what the term "to orbit" means in this connection. Already with electrons, the field of application of the intuitive concepts of the older physics, such as position, velocity and energy, has clearly been restricted. Without this restriction, we should not be able to understand the stability of the atoms. We can represent in mathematical formulae what may possibly be happening to the electron in a given experiment, but we can no longer render this statement objective, as a statement about the electron alone. The question of what electrons, or other elementary particles, such as protons and neutrons, may consist of, leads still further into unintuitable abstraction. Can these entities be further divided, into still smaller building-blocks, or are they truly indivisible ultimates, as in the atomistic philosophy of Democritus? The answer has been given in the last twenty years by the giant accelerators: if two elementary particles are made to collide at high energy, many parts may arise in the process of disintegration, but the pieces are not necessarily smaller than the particle fragmented. In reality, it is a question of producing new elementary particles from the kinetic energy of the colliding partner. The concept of division has thus lost its meaning, and so, too, has the concept of the smallest particle. If energy turns into matter, a possibility already previously envisaged in the theory of relativity, this comes about through its taking the form of elementary particles. This form appears in the mathematical description as the representation of a group of transformations, such as rotation in space, or the Lorentz transformation; the elementary particle is characterized, therefore, by its symmetry properties under the transformations of the group. Such statements, unfortunately, are already very far from intuitable, and hardly intelligible to the mathematically untutored reader. Yet again they make it clear, in the first place, that the road into abstraction does not continue forever, but has a well-defined natural terminus; and in the second, that the Platonic question about reality again arises at the goal. Here, in the form of asking: are these smallest entities real building-blocks of matter, or are they merely the mathematical representations of symmetry groups, whereby matter on the larger scale is constructed?

As already mentioned at the outset, this journey into abstraction has encountered stiff resistance, although clearly no other route to understanding could be proposed. So it is typical that, sooner than renounce the intuitable character of our ideas, the resolute opponents of the journey should have preferred to take refuge in those antinomies of the infinite, already discussed by Kant, which prescribe the bounds to our understanding. Thus in earlier times, for example, when dealing with the problem of inheritance, the naive idea arose, that within an apple seed there is again an invisibly small apple-tree, that this tree in turn bears blossom and fruit, that within this fruit there are again still smaller apple trees growing, and so on, *ad infinitum*. Similarly naive ideas are put forward even today by some specialists in elementary particle physics; for instance, that protons are made up of still smaller entities, the so-called quark particles, that the latter are again composed of still smaller particles, for which the name parton has been suggested, and so on, *ad infinitum*. Obviously, our mind is fighting with all its strength against recognizing that the road to understanding passes out of the intuitable, but yet leads to its goal after a finite number of steps. Such a defense is coupled, perhaps, with the fear that, on reaching the goal, science itself might come to an end. Though that would be simply an error. For it is always just particular fields of a science that can be closed off—mechanics, electricity and heat-theory may be cited as examples—but never the whole science. By intuitability in this connection is meant the world of ideas imposed upon us by everyday experience, which has thus formed since childhood the basis for our finding ourselves at home in this world at all. It is very understandable that we are ready to sacrifice this intuitability only with extreme reluctance. We may perhaps say with some exaggeration, that at the end of the journey we no longer find life or the world any longer, though we do find understanding and clarity concerning the Ideas whereby the world is made.

But since intuitability, too, must always appertain to any living understanding, the end of the journey produces a relentless clarity about the boundaries that are set to rational understanding as such—an impasse that is also familiar in modern psychology. In the final pages of his book, Erich Heller speaks of the tribulation and loneliness of the goal, and concludes with a quotation from Wittgenstein's philosophical reflections, which sounds like a cry of despair: "What is your aim in philosophy? To show the fly the way out of the fly-bottle." And Erich Heller adds: "Here there is no way out." Perhaps we should contrast this quotation with a remark of the physicist and philosopher Niels Bohr, in which light and darkness are distributed in equal measure: "The meaning of life is this, that it has no meaning to say that life has no meaning." Here, too, the boundaries of rational thought are defined with relentless precision, but with this there also goes the idea that every end is at once a beginning. The fact that in science the goal can be reached after a finite number of steps, arouses hope that from hence a new and more ample kind of thinking might originate, though in our own time it can be more readily anticipated than described.

Epilogue

by Hans-Peter Dürr

At the end of 1970, Werner Heisenberg retired from his positions as Superintending Director of the Institute for Physics and Astrophysics, and as Director of the Sub-Institute for Physics in Munich. In 1942 he had taken on the scientific headship of this Institute, the former Kaiser Wilhelm Institute for Physics, established in Berlin-Dahlem in 1917. After the war ended, he set up and led the Max Planck Institute for Physics, which at that time was still housed in the buildings of the Aerodynamical Experimental Establishment in Göttingen.

On the occasion of the ceremonial transfer of office, on December 17, 1970, I myself, as provisionally appointed Head of the Institute for Physics, had the agreeable duty and great pleasure of thanking Dr. Heisenberg for his almost thirty years of distinguished work as scientific leader and director of the institute. In the name of the institute, I could express to him our profound gratitude, that through his generosity and vital interest in the scientific life, he had created in the institute an atmosphere in which learning and research could be carried on in so intensive and fruitful a manner; and that his personal influence should have furnished us with insights that have profoundly affected and altered our own lives. I went on in the following words:

> It would far outstep the bounds of this occasion, were I to attempt a proper estimate of the personality and scientific achievement of Werner Heisenberg. It would mean going into the 200-odd scientific papers and the many books that he has produced in what is now almost fifty years of research activity. I can forego such a detailed review with a good conscience, since very much of it has by now become the common property of all physicists. Much has pushed out beyond the narrow field of physics, and has enduringly altered our understanding of the world, and of ourselves. I am thinking in particular here of Heisenberg's decisive contribution to the founding of quantum mechanics. It strikes us younger people as almost an anachronism, to stand facing in person the originator of these ideas, since in our estimation, Heisenberg himself, and the beginnings of quantum mechanics, have already become part of history.
>
> Heisenberg's influence on physics was profound and many-sided. Arnold Sommerfeld, his teacher in Munich, wanted him first of all to learn a solid trade, and set him a problem from the theory of turbulence in flowing liquids. But the fascination of atomic physics soon became stronger, and gave birth in quick succession to all those writings which eventually led to quantum mechanics, as we nowadays learn it in the schoolroom. Yet even after this great achievement, Heisenberg could always be found at the most interesting places: in 1928 he was able to give the quantum-theoretical interpretation of ferromagnetism; with Wolfgang Pauli he began, in 1929, the quantization of field-theories; in 1932 and the years following he wrote

basic works on the structure of atomic nuclei and their forces; in 1936 he was working on cosmic radiation in the upper atmosphere, and its properties at high energies; during the war years he did calculations on nuclear reactors, and later attempted a theory of superconductivity. In 1950, Heisenberg turned once more to a basic problem, the attempt at a unified quantum theory of the smallest building-blocks of matter, the elementary particles. This fundamental theory still ranks as his main interest even today.

From the standpoint of these last twenty years of research, which have been almost wholly devoted to this unified theory of matter, Heisenberg's earlier work appears as a sort of preparation for this comprehensive and difficult task.

His concern with the turbulence properties of liquids may have precipitated his realization that nonlinear systems, even on small perturbations, can lead to quite unexpected forms of solution, and that quite specific dimensionless numbers are characteristic in describing them. In his unified theory of elementary particles, the nonlinear spinor theory, a special importance attaches to precisely these two points of view. From his intimate knowledge of atomic physics comes a rich store of experience on quantum-mechanical many-body systems, and the important insight that even an essentially simple dynamics can lead to forms of phenomena that are exceedingly complex and impossible to unravel in practice. From this experience he draws assurance that even the complex phenomena of elementary particle physics do not have to stand in contradiction to a simple dynamic set of laws. His encounter with the phenomenon of high-energy radiation in the upper atmosphere led him, quite early on, into denying the elementary nature of elementary particles, long before this became generally evident through the enormous increase in new particles, resulting from experiments made with the great modern accelerators. The quantum field theory, initiated with Pauli in 1929, forms the appropriate point of methodological contact with his unified field theory. In the limited validity of the notion of an elementary particle, he perceives the possibility of a limited validity to quantum theory, and hence a saving escape from earlier difficulties. An indefinite metric in the quantum-mechanical state-space provides him with the formal handle for this. His theory of ferromagnetism finds a striking new application as model for the asymmetrical vacuum-state in modern elementary particle theory.

In this hour of celebration, I cannot and do not wish to enter further upon these many interesting questions and issues. They have found their due embodiment in the scientific literature, and should receive notice in another connection.

I would like instead to dwell upon a different point, which has been of outstanding significance to myself, as to others, who have been fortunate enough to work with Werner Heisenberg. Heisenberg has shown us in striking fashion what it means to seek, to enquire, to understand and to know: that it means to dedicate oneself to a task without reservation; to reach toward the difficulties by laborious detailed work; to try to separate the essential from the inessential; and not to succumb to the temptation of sacrificing content to form. Heisenberg is not afraid of getting his hands dirty, of himself grubbing in the soil for fertile roots, and shifting obstructive boulders out of the way. He knows the country as a farmer knows the acreage he has ploughed and planted for years, who can tell where something is growing, and where not. He has taught us that you cannot solve complex problems by generalizing them, but rather by making them concrete—by embedding them, that is, in a world of ideas that is currently and immediately intelligible to us on the strength of our previous experience. In this accustomed world of thought, the novel can be far more simply accommodated, and our limited imagination finds it much easier to light on fruitful ideas. Heisenberg has spared no pains in sounding out every new scientific advance with a view to its essential significance; he has not been content to derive a result with mathematical correctness, but always tries to think it over closely from all angles, to grasp it fully in thought, to understand it—which means to order it within a conceptual universe that at once bespeaks the logical universe of mathematics. We have always been deeply impressed by his unwavering and concentrated purposefulness in following out his ideas, his diligence in pursuing detailed investigations, his optimism and energy in attacking new and difficult tasks. But his purposefulness is not rigidity; it is coupled with that peculiar sensibility which reacts sensitively to discrepancies that give warning of profound changes. He is endowed in a marked degree with the capacity for leaving questions open, for fitting them at first, without impatience, only crudely and vaguely into the framework of ideas. Novel thoughts, whose treatment seems not yet immediately accessible, he is happy to leave dangling, in order to protect them against prejudices, which spring all too easily from our want of understanding, and to guard them against over hasty criticism, which is often a mere expression of our limited imagination in thinking the unusual. Such thoughts should first ripen before one weeds them over with harsh criticism. He has shown us how, in this way, one can still find solutions, even in a situation where there seem to be no prospects in sight.

In the many scientific conversations we have had with him, the distinction between teacher and student has vanished. We have been immersed in the problem at issue, trying to grasp it, to pass it on to the other; we have talked about it, stumblingly and unintelligibly, and in spite of that the other has understood. We have criticized severely, and yet could never give offence. The common problem came foremost, and the wish to grasp and resolve it. A false indication could induce depression, an 11 P.M. telephone call clear up the trouble with the cheerful news: "It'll work!"

You have taught us, most dear and honored Dr. Heisenberg, that science can be a tremendously exciting thing, if one is ready to give it one's utmost. But you have also made us realize that science, when pursued together, can lead to the most delightful of human contacts. We should like most especially to thank you for that.

November 1, 1976

THE PRINCETON SCIENCE LIBRARY

Edwin Abbott Abbott — *Flatland: A Romance in Many Dimensions*
With a new introduction by Thomas Banchoff

Philip Ball — *Designing the Molecular World: Chemistry at the Frontier*

Friedrich G. Barth — *Insects and Flowers: The Biology of a Partnership*
Updated by the author

Marston Bates — *The Nature of Natural History*
With a new introduction by Henry Horn

John Bonner — *The Evolution of Culture in Animals*

A. J. Cain — *Animal Species and Their Evolution*
With a new afterword by the author

Jean-Pierre Changeux — *Neuronal Man: The Biology of Mind*
With a new preface by Vernon B. Mountcastle

Paul Colinvaux — *Why Big Fierce Animals Are Rare*

Peter J. Collings — *Liquid Crystals: Nature's Delicate Phase of Matter*

Pierre Duhem — *The Aim and Structure of Physical Theory*
With a new introduction by Jules Vuillemin

Manfred Eigen & Ruthild Winkler — *Laws of the Game: How the Principles of NatureGovern Chance*

Albert Einstein — *The Meaning of Relativity*
Fifth edition

Niles Eldredge — *The Miner's Canary: Unraveling the Mysteries of Extinction*

Niles Eldredge — *Time Frames: The Evolution of Punctuated Equilibria*

Claus Emmeche — *The Garden in the Machine: The Emerging Science of Artificial Life*

Richard P. Feynman	*QED: The Strange Theory of Light*
Solomon W. Golomb	*Polyominoes: Puzzles, Patterns, Problems, and Packings* Revised and expanded second edition
J. E. Gordon	*The New Science of Strong Materials, or Why You Don't Fall through the Floor*
Richard L. Gregory	*Eye and Brain: The Psychology of Seeing* Fifth edition
Jacques Hadamard	*The Mathematician's Mind: The Psychology of Invention in the Mathematical Field* With a new preface by P. N. Johnson-Laird
J.B.S. Haldane	*The Causes of Evolution* With a new preface and afterword by Egbert G. Leigh
Werner Heisenberg	*Encounters with Einstein, and Other Essays on People, Places, and Particles*
François Jacob	*The Logic of Life: A History of Heredity*
Rudolf Kippenhahn	*100 Billion Suns: The Birth, Life, and Death of the Stars* With a new afterword by the author
Hans Lauwerier	*Fractals: Endlessly Repeated Geometrical Figures*
Laurence A. Marschall	*The Supernova Story* With a new preface and epilogue by the author
Helmut Mayr	*A Guide to Fossils*
John Napier	*Hands* Revised by Russell H. Tuttle
J. Robert Oppenheimer	*Atom and Void: Essays on Science and Community* With a preface by Freeman J. Dyson
John Polkinghorne	*The Quantum World*
G. Polya	*How to Solve It: A New Aspect of Mathematical Method*

Hans Rademacher & Otto Toeplitz	*The Enjoyment of Math*
Hazel Rossotti	*Colour, or Why the World Isn't Grey*
Rudy Rucker	*Infinity and the Mind: The Science and Philosophy of the Infinite* With a new preface by the author
David Ruelle	*Chance and Chaos*
Henry Stommel	*A View of the Sea: A Discussion between a Chief Engineer and an Oceanographer about the Machinery of the Ocean Circulation*
Geerat J. Vermeij	*A Natural History of Shells*
Hermann Weyl	*Symmetry*
George C. Williams	*Adaptation and Natural Selection* With a new preface by the author
J. B. Zirker	*Total Eclipses of the Sun* Expanded edition

www.ingramcontent.com/pod-product-compliance
Ingram Content Group UK Ltd.
Pitfield, Milton Keynes, MK11 3LW, UK
UKHW022101160125
453814UK00010B/314